ATTRACTING SUSTAINABLE INVESTMENT

This book is a practitioner's guide to sustainable development, laying out strategies for attracting investment for communities and their partners.

It proposes an innovative Sustainable Development Proposition (SDP) decision-making tool based on a propositional calculus that can be used to analyse the sustainability of an infrastructure investment. It draws on environmental sustainability governance data analysis enabling investors to understand the economic indicators, income potential, return on investment, demand and legal compliance, as well as community and social benefits. Identified risks, issues and advantages are managed and monitored, and the SDP guidance can be applied to improve the prospects of the project in order to attract investment.

Sustainable Community Investment Indicators (SCIIs™) have been developed to assist with attracting investment and monitoring feedback on infrastructure projects, designed by the author for remote rural and indigenous communities – in response to current industry tools that are designed for urban environments. The book includes a broad range of real-world and hypothetical case studies in agricultural and indigenous areas in South America, Europe, Africa, Asia, Australia and the Pacific.

Taking a diverse economies approach, these industry tools can be adapted to allow for enterprise design with unique communities. This book provides sustainable development practitioners, including government agencies, financiers, developers, lawyers and engineers, with a positive, practical guide to addressing and overcoming global issues with local- and community-based solutions and funding options.

Saskia Vanderbent is a lawyer based on the Gold Coast, in sunny Queensland, Australia. She holds a multi-disciplinary PhD from the faculty of Engineering and a Juris Doctor in Law.

"Saskia's study of what works when investing in sustainable energy in remote and Indigenous communities will be a great boon to both those communities and potential investors. The text makes fascinating reading. Her conceptual tool, the Sustainable Development Proposition, makes application of that knowledge a little easier in practice."

Stephen Keim *SC, Barrister-at-Law,*
recipient of the Law Council of Australia's 2020 President's Award
and recipient of the Human Rights Medal, 2009,
by the Australian Human Rights Commission

"The tools in this book will ensure that your return on investment goes beyond a monetary return. You will have a sound basis to expect that your investments, and the partnerships that are formed with communities, are building capacity at the local level, and enhancing connectivity and opportunities with the broader economy in a way that is, dare I say, sustainable in perpetuity."

Craig Cowled, *Engineer, Researcher, Educator, Worimi man*

"Saskia Vanderbent has provided a comprehensive insight into the diverse energy challenges being confronted globally and how communities are moving to address them. This prescient perspective takes its currency in the present circumstances facing the world."

Allan Fife *OAM, Chief Investment Officer, Fife Capital*

ATTRACTING SUSTAINABLE INVESTMENT

A Professional Guide

Saskia Vanderbent

LONDON AND NEW YORK

Cover image: Pixabay from Pexels

First published 2023
by Routledge
4 Park Square, Milton Park, Abingdon, Oxon OX14 4RN

and by Routledge
605 Third Avenue, New York, NY 10158

Routledge is an imprint of the Taylor & Francis Group, an informa business

British Library Cataloguing-in-Publication Data
A catalogue record for this book is available from the British Library

Library of Congress Cataloging-in-Publication Data
Names: Vanderbent, Saskia, author.
Title: Attracting sustainable investment : a professional guide / Saskia Vanderbent.
Description: Abingdon, Oxon ; New York, NY : Routledge, 2023. |
Includes bibliographical references and index.
Identifiers: LCCN 2022033812 | ISBN 9781032349633 (hardback) |
ISBN 9781032349596 (paperback) | ISBN 9781003324669 (ebook)
Subjects: LCSH: Renewable energy sources–Finance. |
Investments–Environmental aspects. | Infrastructure (Economics)–Finance. |
Sustainable development–Finance.
Classification: LCC HD9502.A2 V36 2023 |
DDC 333.79/4–dc23/eng/20220728
LC record available at https://lccn.loc.gov/2022033812

ISBN: 978-1-032-34963-3 (hbk)
ISBN: 978-1-032-34959-6 (pbk)
ISBN: 978-1-003-32466-9 (ebk)

DOI: 10.4324/9781003324669

Typeset in Bembo
by Newgen Publishing UK

CONTENTS

FIGURES

TABLES

AUTHOR BIOGRAPHY

Saskia Vanderbent is a lawyer based on the Gold Coast, in sunny Queensland, Australia. She holds a PhD in Engineering and a Juris Doctor in Law. She has designed an ESG index, SCIIs™ (Sustainable Community Investment Indicators)™. She also developed a decision-making tool using propositional calculus, the SDP (Sustainable Development Proposition). She provides advice to governments and multinational companies.

Prior to opening her law firm, she was a senior litigator for Crown Law, the Department of Justice and Attorney General in the Native Title and Resources litigation team. She is admitted to appear in the High Court of Australia and can seek leave to appear in the International Court in the Haag. She has experience in the Federal Court, Land Court, Supreme Court and Planning and Environment Court.

She has also designed and coordinated several university law and property economics courses.

FOREWORD

Life is a gift granted to all of us by the Earth. It is a gift that will keep giving in perpetuity if we are wise and care for the Earth. Wisdom about the Earth, and knowledge of the complex interdependencies of its life systems, has been collected by indigenous peoples around the world who have a long and deep connection with their local area – their country. Progress, in all its forms – including scientific, technological, economic, political and social progress – has tended to push aside and suppress local wisdom and knowledge in favour of *universal, one-size-fits-all* solutions to problems. While this approach might work for the majority, who live in cities with ready access to high-quality infrastructure, it doesn't always work for those on the margins of the developed world – that is, those in rural/remote areas and those in developing countries – and it certainly hasn't worked for the environment.

The local wisdom and knowledge of indigenous peoples can help us understand Earth's complex systems and guide us in our walk on this Earth. I would like to share a dreaming story of my own people, the Worimi. Since time immemorial, the Worimi have lived on the shores of a large and beautiful bay, nowadays called Port Stephens, which is where this dreaming story comes from. Dolphins would play in the warm waters of the bay where they had a plentiful supply of fish to keep their tummies full. One dolphin, named Guparr, had a realisation that the fish would go extinct if she kept eating them without restraint. She enlisted the aid of another dolphin, named Wubaray, and together they kicked up some rocks from the bottom of the bay with their snouts and powerful tails to form islands and hiding places for the fish. The great spirit, Baiame, saw what Guparr and Wubaray had done to protect the fish and he rewarded the dolphins for their selfless actions by transforming them into the first human beings. This story teaches us that it is our primary function, as humans, to protect the environment that supports us.

I met Saskia several years ago when she approached me about her desire to contribute in a practical way to improve access to critical infrastructure in rural and remote indigenous communities. Saskia had become aware of the lack of infrastructure in these communities while working as a researcher for the courts and as a lawyer in native title. I was very interested in Saskia's proposal and offered to be a sounding board for her research and I am glad for the opportunity to be a witness to the development of the ideas in this book. It is my hope that Saskia's work will help to connect capital with communities to improve the standard of living and support the economic development of these communities in a sustainable and responsible way.

In this book, Dr Saskia Vanderbent presents a flexible tool, and some indicators, that can be used to attract investment in critical infrastructure for people who have, to date, been poorly served by progress. Saskia's method is adaptive to the specific needs and values of a community. Importantly, Saskia's approach is aligned with the principles of sustainability. That is to say, the tools in this book can be used to attract investments in infrastructure that are sustainable in terms of the environment, the technology being deployed, the social and governance structures, and economics.

Craig Cowled

Engineer, Researcher, Educator, Worimi man.

PREFACE

Early on in my career as a lawyer, I worked as a researcher for the Land Court and Aboriginal Land Tribunals, as a Registrar in the Supreme Court and then subsequently as a senior lawyer in the Native Title and Resources Division of Crown Law, appearing in the Federal Court. During that time, I became interested in issues of economic justice as they related to remote communities. Having studied economic justice in my philosophy degree as an undergraduate, I was interested in how theories of justice applied in real-life contexts. The court participants I assisted came from diverse pastoral, Australian Indigenous, mining, government and infrastructure groups. I became proficient at negotiating with multiple participants from these diverse communities who all shared common resources. Many of these diverse participants were concerned about the sustainability of the resources in these remote locations and lack of infrastructure supply.

Although their interests often differed, commonality could be found. Whether a farmer, an Indigenous elder or a mining company, each participant wanted the water, land and energy supply to be sustainable. This experience sparked my interest in sustainable energy security for remote communities. I particularly saw a gap in energy justice for remote farming and Indigenous communities in regional Australia. Due to the vast distances, poles and wires cannot be built to all regions and many communities were still reliant on diesel generators, an unaffordable energy source that is heavily subsidised. The increasing affordability of small-scale solar and wind generation and the research conducted by former Chief Scientist, Dr Finkel, in 2017 into the advantages of decentralised off-grid systems for remote communities inspired this research.

It was apparent that the gap in remote community infrastructure supply was one of access to legal and financial services or "soft technology" and not just the physical technological infrastructure. I focused my research questions on closing that gap. I soon discovered that the issue of infrastructure supply could not be solved by mere provision of public government grants or private financial institutional funding because many projects funded this way in the past had become defunct due to unsustainable models. For example, where infrastructure was unable to be fixed, maintained or upgraded under a grant-only model, or where funding was unavailable from institutional investors due to the underlying security assets required for loans or the high-risk nature of the investments. The research discovered a variety of structures and methods that have been used successfully to raise funds and attract investment in regional energy projects.

While researching the supply of sustainable solar infrastructure to cities to see how these models might be adapted to remote communities, I discovered that funding bodies and private investors were using a new concept called "smart city indicators" which were a set of tailor-made indicators used to assess infrastructure gaps and to inform investment decisions on infrastructure builds for large cities. It was from this research that the idea was born to develop a set of remote community investment indicators designed to attract investment in areas where there was an infrastructure gap.

I realised that although all remote communities had similar needs of energy security, energy justice and sustainable management of resources, a diverse economies perspective was required. Diverse economic thinking and community-based economic action were needed to close the gap in infrastructure supply. This is particularly the case in energy and infrastructure. Mining, pastoral and indigenous communities often share overlapping interests over the same parcels of land. As there is a diverse range of needs between these communities, there is a responsibility to apply context-based, diverse perspectives in the enterprise design phase of an infrastructure project. A diverse economies perspective enables recognition of the multitude of ways that individuals and communities can engage, participate and contribute to economic, social and environmental aspects of a project. This approach allows for responsive, place-appropriate generation of a diverse range of economic outputs and inputs. It allows for free choice in legal structures such as companies, trusts, associations and cooperatives and allows individuals to choose their ownership and voting rights, fund-raising options, trading activities, tax status and so on. It relies on community involvement in making those enterprise design choices.[1]

This industry guide analyses and develops a diverse range of economic and legal structures beyond the traditional government grant based or financial institutional based investment models. It provides a cost-benefit analysis and recommendations in relation to those models. Further, this industry guide provides a set of community indicators, the Sustainable Community Investment Indicators designed as a decision-making tool for investment in remote community infrastructure and a set of propositions, the Sustainable Development Propositions, that inform those indicators by reviewing what it means to be a "sustainable development project."

Note

1 See Gibson (2018, 613–618) for more information on enterprise design theory.

ACKNOWLEDGEMENTS

I would like to offer my sincere thanks to my family for their support during my journey.

I would like to acknowledge the Queensland University of Technology and the mentoring from my supervisors. This book was made possible by the guidance of my supervisor Dr Craig Cowled. Dr Cowled is an Indigenous Australian and is passionate about environmental issues.

I would like to thank my external supervisor Dr Allan Fife (University of Western Sydney) for his valuable feedback and support throughout my journey. Dr Fife's knowledge of property economics and his inspirational work in industry has guided me throughout this research.

I would like to acknowledge the support from the Queensland University of Technology's Office of Research and my colleagues at the School of Civil and Environmental Engineering.

I would like to acknowledge and express my gratitude to the Commonwealth of Australia for providing a research scholarship and the Queensland University of Technology for providing a write-up scholarship. This research was supported by an Australian Government Research Training Program Scholarship.

A thank you to my past mentors and colleagues from the Queensland Land Court, Queensland Courts, Aboriginal Land Tribunals, Crown Law, Bond University, the University of Queensland and the Australian National University.

A big thank you to the volunteer panel members who provided feedback and assessment of this research, Stephen Keim SC of the Queensland Bar and Associate Professor Geoff Walker.

ABBREVIATIONS

ABS	Australian Bureau of Statistics
ACCC	Australian Competition and Consumer Commission
ACT	Australian Capital Territory
AEMC	Australian Energy Market Commission
AEMO	Australian Energy Market Operator
AER	Australian Energy Regulator
CAPEX	Capital expenditure
CDEP	Community Development Employment Project
CfAT	current name for Centre for Appropriate Technology Limited
CAT	former name for Centre for Appropriate Technology Limited
COAG	Council of Australian Governments
CSIRO	the Commonwealth Scientific and Industrial Research Organisation
DER	Distributed energy resources
DOGIT	Deed of Grant in Trust
FiTs	Feed-in tariffs
GDP	gross domestic product
HDI	Human Development Index
IBA	Indigenous Business Australia
ICEs	internal combustion engines
ICT	Information and Communications Technology
IEA	International Energy Agency
ILM	indigenous land management
ILUAs	Indigenous Land Use Agreements
IPCC	Intergovernmental Panel on Climate Change
IRR	internal rate of return
MOU	Memorandum of Understanding
NEM	National Electricity Market

NFF	National Federal Farmers Association
NFP	not-for-profit organisation or non-profit
NGO	non-government organisation
NIRA	National Indigenous Reform Agreement (Closing the Gap)
NSW	New South Wales
NTRBs	Native Title Representative Bodies
OPEX	Operating expenditure
PERC	Passivated Emitter and Rear Contact solar cell
PPA	power purchase agreement
PV	photovoltaic
Qld	Queensland
QUT	Queensland University of Technology
REC	Renewable Energy Certificate
REDP	Renewable energy development programme
REITs	renewable energy certificates
RNTB	Registered Native Title Body
ROI	return on investment
SCIIs™	Sustainable Community Investment Indicators
SDP	Sustainable Development Proposition
STEM	Science Technology Engineering Maths
UNSW	University of New South Wales
VPP	Virtual power plant
WPO	waste plastic oil

1

THE SUSTAINABLE INVESTMENT MARKET

Introduction

Today, societal and planetary health is at the forefront of our public discourse. Humanity is currently experiencing significant feedback from our environment. We have internal and external feedback functions that we monitor to improve what we could do better in the future. Recognising and balancing those functions is how we both survive and fulfil our obligation as sentient custodians of the Earth, our planetary home. As we celebrate the achievements of the past, we must build on those achievements for future generations. The developed world has been thriving over the last few generations, keeping peace, ensuring abundant food and clean water, trading natural resources and increasing environmental awareness. Human rights have been improving and the overall standard of living has been increasing as technology and resource management improve.

However, remote rural and indigenous communities globally have not benefited to the same degree. This is largely due to the vast distances between remote communities and urban centres. This infrastructure gap causes decreased standards of living.

There are 1.2 billion people worldwide who lack access to electricity and 2.8 billion who rely on biomass for cooking and heating.[1] These communities lack the basic living standards that urban centres take for granted. In rural and remote communities, renewable energy infrastructure can increase standards of living and provide economic growth.[2] For example, in Latin America alone there are 17 million people without electricity due to the isolated locations such as the Amazon which have made it difficult to extend the grid to those areas.[3]

Many remote communities are reliant on heavily subsidised diesel generator systems. This causes an unaffordable and unreliable electricity supply.[4]

This guide poses the question: What strategies can be used to attract investment in sustainable energy infrastructure in rural and indigenous communities?

DOI: 10.4324/9781003324669-1

The question is answered through the exploration and development of sustainable investment models that can be applied in regional and indigenous communities to improve energy security and affordability for those communities. The author has designed two tools for attracting investment: A decision-making tool, the Sustainable Development Proposition (SDP), and a decision-making index, the Sustainable Community Investment Indicators (SCIIs™).

This guide presents the following:

1. A cost-benefit analysis of renewable energy and sustainable development case studies which are located in rural and indigenous communities. The cost-benefit analysis evaluates the financial, legal and technological structures in community case studies and analyses them for sustainability and their attractiveness to investors.
2. The most sustainable and attractive investment strategies are distinguished, selected and recommended for use by industry professionals.
3. Novel hybrid investment structures are also developed out of this analysis and are identified in the recommendations.
4. This guide reviews and analyses the question, what is a Sustainable System?
5. This question is answered with the SDP tool which was developed to assist communities in designing a Sustainable System .

   The SDP analyses investment opportunities to test their level of sustainability. Sustainability is defined in the proposition to consist of the following four key objectives:

   - Economically self-sufficient and sustainable, that is, they are independently profitable without needing ongoing government funding;
   - Environmentally sustainable;
   - Governed sustainably; and
   - Technologically sustainable.

The SDP is made up of nine propositions that contribute to a Sustainable System. Each of those propositions falls within one of the four above categories. Where those propositions are satisfied, a Sustainable System or sustainable community exists, enjoying an increased standard of living and social benefits, such as energy reliability and affordability, health benefits, education and employment opportunities.

6. The SCIIs™ are an innovative set of indicators that are specifically designed for attracting investment in and monitoring outcomes of infrastructure built in remote and regional communities as opposed to urban areas where indicators have been previously focused.

The SCIIs™ are a product of the SDP and the results of the case study analysis.

These two interdependent systems are designed to be used within an expert's own decision-making process and are adaptable for implementation in a variety of diverse communities.

This work is significant not only because there is currently a gap in energy security for remote communities but also because of the opportunities that addressing this gap can provide for those communities. By creating a sustainable infrastructure project, local employment opportunities, economic growth and contribution to a more sustainable management of resources can be achieved by communities and their investor partners.

This guide challenges the current reliance on traditional funding methods of publicly funded government grants and private institutional investment and recognises that while a government regulatory system is needed in the property development and electricity marketplaces, that system supports an economically diverse plethora of sustainable investment models to meet the needs of the diverse communities in remote areas.

Although all remote communities have similar needs of energy security, energy justice and sustainable management of resources, not all communities are the same and a diverse economies perspective is required. Diverse economic thinking and community-based economic action are needed to close the gap in infrastructure supply. Consequently, this guide analyses and proposes a diverse range of economic investment and legal structures beyond the traditional government grant based or financial institutional based investment models.

Globally, the diversity of choice within the free markets is what attracts the vast range of investors to community projects in western countries. The detailed case study analysis in this industry guide demonstrates the importance of context-specific enterprise design on community autonomy and the sustainability of projects. Participation by communities needs to be adaptable in order to generate participation that is significant, sustained and diverse.

To this end, each case study in this guide explores context-specific engagement, economic arrangements, legal structure practices, technology choices and sustainable outcomes related to those context-specific choices. It was found that community participation in enterprise design increased local investment and involvement, creation of local relationships and employment, and increased skills in the communities. Overall, a range of social, economic and environmental impacts was reviewed across a variety of diverse community projects.

The Human Development Index (HDI) was developed by the International Energy Agency (IEA) to track how energy improves the socio-economic condition in developing countries.[5] This improvement in the standard of living can be indexed but it can also be shown in specific examples. For instance, in remote Australian communities, supply is usually provided by the power card system or diesel generator micro grids. When power cards run out of credit, the power supply to the house is cut off.[6] This system is used throughout the Northern Territory and North Queensland.[7] The reported implications of this include the inability to refrigerate medication or fresh food, inability for children to do homework in the evening, impact on hygiene and washing of clothes, lower school attendance, decrease in physical health, decrease in employment and lack of air conditioning or heating affecting well-being.[8] It was reported that many people choose to sleep

FIGURE 1.1 Wind turbines on a large agricultural holding. Photo by Alexey Komissarov from Pexels.

outside due to heat and lack of air conditioning, thereby increasing snake and dog bites and women being harmed.[9]

This book is a guide that focuses on strategies for attracting investment in sustainable energy infrastructure for those communities. Government agencies, institutional investors and not-for-profit organisations have signalled that they will be increasing investment in sustainable energy infrastructure for remote communities.

It is expected that globally 485 million people in isolated regions globally will gain access to electricity by 2030 through decentralised non-network renewable energy solutions such as solar photovoltaic (PV) and micro-grid solutions.[10] However, community energy products often fail because of lack of support from government, lack of governance structures that would attract private investment, lack of access to financial products and changing policies.[11] Projects also fail because of lack of community involvement, trained people and ongoing funds to ensure continued operation and maintenance.[12] This guide focuses on case studies that demonstrate successful renewable energy projects in remote regions. A cost-benefit analysis is undertaken to discover the successful funding and governance structures that attract investment and support sustainable projects.

A decision-making tool (the SCIIs™) has been developed by the author to assist communities in the pre-construction phase of their infrastructure projects. An advantage of the SCIIs™ is that they are designed to be flexible and to work across a variety of community contexts. Democracy requires that alternative options must be provided to nations and their policy makers as policy options and choices, not externally applied homogenised laws.[13] It is this vein that the research into guidance for rural and indigenous communities has been conducted, to provide choices.[14] There are different tools that can be used to provide this sustainability analysis to communities. Leaders can implement or revisit sustainability assessments within their own policy and institutional contexts.[15] This is in keeping with underlying democratic principles.

A property economics approach can evaluate models to supply sustainable infrastructure to remote communities. This approach can overcome the various barriers to infrastructure supply that currently exist. This guide recommends new investment models that can attract investment in those communities where there is currently a gap and ensure that those investments are sustainable.

These investment models are evaluated, contrasted, differentiated and selected through the cost-benefit analysis of renewable energy community case studies.

The SDP is an innovative decision-making tool based on propositional calculus that can be used to analyse the sustainability of an infrastructure investment. The tool can use environmental sustainability governance data analysis to identify stress points as well as issues and advantages of a proposed investment. Investors will be informed in relation to economic indicators, income potential, return on investment, demand and legal compliance as well as legal and social governance. Any risks or advantages identified are managed and monitored in a templated system which when applied can improve the prospects of the project in order to attract investment.

The SCIIs™ have been developed to assist with attracting investment in an infrastructure project and also to monitor and give feedback on projects so that changes can be made where needed.

As shown in Figure 1.2 these two systems, SDP and SCIIs™, are designed to be applied across a variety of diverse remote communities and investments.

The SDP and SCIIs™ in a renewable energy project

Figure 1.2 shows the relationship between the SDP and the SCIIs™ and how this can be applied to a renewable energy project. For example, each proposition relates to indicators that will promote benefits and increased standards of living for communities. The SCIIs™ are benefits that can be achieved in a renewable energy project. They are the outcomes of the Sustainable System .

The sustainable investment market

The future economy will be centred on products and services that are less energy intensive, more water efficient and circular, based on closed-loop systems. Solar power in particular attracts significant investment. A joint international investment plan with a goal to invest $10 trillion in the next ten years was launched in October

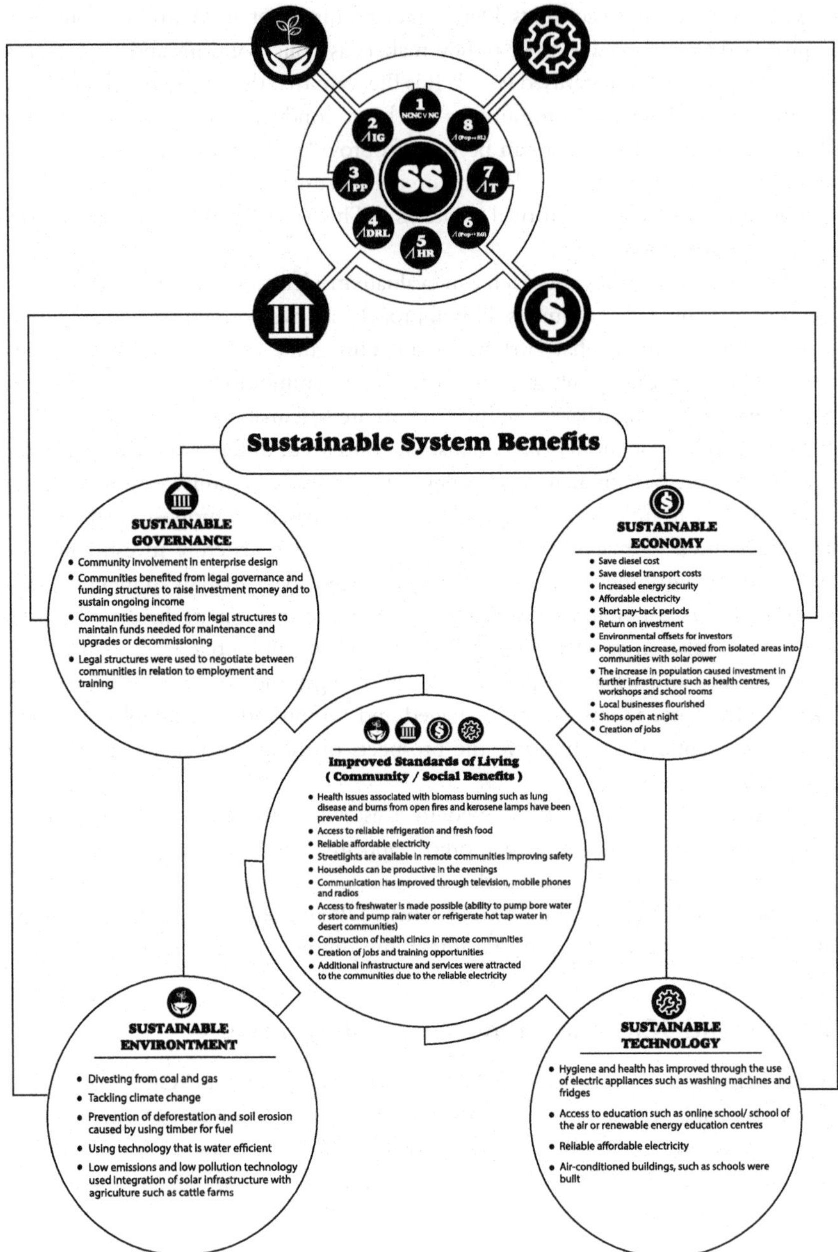

FIGURE 1.2 Relationship between the Sustainable Development Proposition (SDP) and the Sustainable Community Investment Indicators (SCIIs™).

2020 by a consortium made up of the Australian Energy Market Operator (AEMO), the National Grid Electricity System Operator, UK, California Independent System Operator (CAISO), the Electric Reliability Council of Texas (ERCOT), Ireland's System Operator (EirGrid) and Denmark's System Operator (Energinet).[16]

The Reserve Bank of Australia stated in 2020 that coal-fired generation capacity will be retired over the next couple of decades in Australia, to be replaced by distributed energy sources and large-scale renewable energy generators, supported by energy storage.[17] Since then the case for clean coal has also been explored.

In December 2019, the European Commission launched the European Green Deal Investment Plan (Sustainable Europe Investment Plan) which aims to fund at least 1 trillion euros of globally sourced and publically funded sustainable investment between 2021 and 2030.[18]

The United Nations Environment Program reported that millions of people are employed in renewable energy sector globally.[19] Australia has an advantage in the renewable energy industry due to its history of scientific innovation, academic excellence and access to trade. For instance, 50% of all silicon solar panels incorporate technology invented in Australia.[20] This gives Australia the edge in leading the world in the solar industry.

Australia has excellent wind and solar resources, with more solar radiation per square metre than any other continent. Only 0.1% of this radiation needs to be converted into electricity to power the nation[21]. Australia also has great mineral wealth that pertains to battery storage and solar panel manufacturing as well as wind turbine manufacturing.[22] Western Australia alone produces 41% of the global supply of hard rock lithium. It is the second largest producer of cobalt and has the largest potential for Class 1 Nickel. It also produces 16% of the world's rare earth metals used in battery technology (the world's second largest producer).[23] The building and construction industry is also advanced and is already competitive in the green building market.[24]

Australia has a diverse range of communities in remote off-grid locations. These communities are poised to transition to renewable energy. The communities are made up of indigenous people, farming and pastoral communities, mining towns and tourist industry areas. Australia is uniquely placed as a developed western democracy with a large number of communities with diverse needs. It is important that a diverse economies perspective is followed to ensure the enterprise design choices of communities are considered and respected. It is also important to ensure that one factor or indicator is not given priority over other sustainable development goals to the detriment of remote communities. For instance, the transition to a renewable energy economy in remote towns must be done responsibly. The goal to reduce carbon in the atmosphere should not outweigh the goal to support living standards and employment in remote communities. The transition should be done in a responsible manner, supporting the local economies of the remote towns such that energy affordability and reliability are not compromised for customers during a transition period. The Australian and Pacific case studies demonstrate that the preferences of the European Union for a globally homogenised transition to a sustainable economy cannot be applied uniformly across every indigenous and remote community globally. Each community has sensitive context-specific needs and considerations that must be taken into account if responsible sustainable development is to occur.

It is for all these reasons that Australia is perfectly placed to attract significant investment in the renewable energy industry. This includes remote and regional areas where infrastructure is due for investment and upgrades. These areas are a prime candidate for investment in cutting-edge renewable energy technology.

When it comes to competing on a global economic scale, energy industries are a significant factor. China's UHV super grid moves 161.5 terawatt hours of hydro, wind and solar annually.[25] In September 2020, China pledged to the United Nations General Assembly to become carbon neutral by 2060.[26] It is this scale of investment that provides an indicator that renewable energy industries are growth industries. Investing in renewable energy can provide a country with independence from petroleum and fossil fuel imports. It is for this reason that renewable energy attracts significant investment by governments and the private sector globally.

Remote communities are often marginalised in the technological race to design sustainable urban infrastructure. The purpose of this guide is to develop strategies for attracting investment in infrastructure for those communities. The models that are illustrated in this guide demonstrate a return on investment; however, there are many other factors such as social benefits that attract investment in communities that are investigated.

Global companies have realised that investment in rural areas is attractive, for example, multinational corporation, Ericsson comments, "reaching the next billion subscribers means expanding to rural off-grid areas." The company has developed highly efficient solar-powered cell towers for use in outlying areas where there is no electric power infrastructure.[27]

It is the goal of many city designers to untether communities from the grid. This is seen as a way to increase the standard of living in urban areas.[28] For example, mobile phone networks and wireless internet connections can be provided to a much higher percentage of people when untethered from the grid. Similarly, communities that are already off-grid can be assisted with off-grid renewable energy infrastructure.[29]

> Local is the perfect scale for smart technology innovation for the same reasons it's been good for policy innovation – it's much easier to engage citizens and identify problems, and impact of new solutions can be seen immediately. [30]

Non-network (off-grid) renewable energy solutions are best suited to the remote regions of Australia in terms of affordability and efficiency of infrastructure builds. As stated by Australia's Chief Scientist, Dr Finkel, this reduces upfront network costs, increases participation in the market and flattens demand peaks. Non-network solutions also create resilience through diversity as they can operate as stand-alone or connected systems when designed to do so.[31] The case studies analysed below demonstrate that this model is often sustainable; economically, environmentally and technologically.

Non-network systems are the perfect starting point for upgrading energy systems. They can encourage innovation, experimentation and diversity, out of which the most technologically, economically and environmentally Sustainable Systems will emerge. Network solutions for remote agricultural communities are also reviewed in the case studies. These systems provide good economic returns and where grid connection is available to supply renewable energy to both the local towns and cities such as Sydney. The case studies in this guide demonstrate solar farms integrated with cattle farms that supply electricity to big purchasers such as the Sydney Opera House and Coles Supermarkets.

Renewable energy solutions are made up of a mix of 100% renewable technologies, such as wind, solar with storage and demand management technology, small-scale hydro and biogas.[32] The case studies have been chosen within this range of renewable energy technology.

This work focuses on remote and indigenous communities. Previously, development indicators have been drafted for metropolitan areas to inform investment decisions by governments and institutional infrastructure investors. Uniquely, this guide creates a set of indicators designed to attract investment from private and community investors for remote farming and indigenous communities. The purpose is to give communities a capital funding alternative to the usual models of government budget allocation and charitable donation. To this end, the set of remote community infrastructure indicators is developed. These indicators are best practice decision-making tools that can be used to attract investment and guide and monitor investment decisions. They can be used to demonstrate the attractiveness of a project to investors.

Government departments and private companies and organisations have begun working towards funding and building solar infrastructure in regional communities. However, more investment is needed. For instance, it is recognised that the majority of community development projects are tied to government assistance, foreign aid and charity.[33] Relying only on foreign investment, government funding and charitable donation will not be adequate to meet the infrastructure needs of all regional communities lacking infrastructure to meet their immediate needs, while at the same time ensuring community development is sustainable with reference to the value of natural capital and maintenance and upgrade of infrastructure in the future.

In relation to economic participation in the electricity market by indigenous communities, COAG (the Council of Australian Governments) has stated that housing infrastructure, including electricity supply, is one of the key targets and indicators for the *National Indigenous Reform Agreement (Closing the Gap) (NIRA)*.[34] Energy supply in remote communities is a key target for closing the gap.

It is further outlined in the NIRA that economic participation by indigenous people should be a part of building the aforementioned housing infrastructure. This would foster economic independence and wealth creation as well as community engagement. "Through this participation, parents and other adults can become effective role models for their families and community."[35]

However, there are still multiple barriers to infrastructure supply in remote communities. The 2018 *Handbook of Contemporary indigenous Architecture* explains the economic barriers to electricity generation in Maori indigenous communities in New Zealand as including "difficulty in obtaining a mortgage to fund construction, as commercial banks are reluctant to use land in indigenous communal title as security over loans."[36] Securing a loan is not easy when there is no source of income. Even if you have freehold tenure, it is very difficult to secure a loan. If it is an investment property, the loan requires a monetary return on investment. Even more difficult is securing a loan where there is no freehold ownership over the property. Ethical investment companies will occasionally take on such risky investments if they are for a good purpose but this is not commonplace.[37]

Other inhibiting factors are the "costs of bringing electricity and fresh- and grey-water services to new rural building sites."[38] Many communities are situated on leasehold or Deed of Grant in Trust Land where lack of freehold ownership is a barrier to funding arrangements. For example, in Cherbourg, Queensland, Aboriginal residents pay rent to the Aboriginal Shire Council if living on Deed of Grant in Trust Land. The Aboriginal Shire Council sets the amount of rent to be paid and the residents do not own the property.[39]

Community development relies on principles, processes and practices. Additionally valuation principles and land tenure play a key role in those processes. For example, when assessing renewable energy models from a property economics perspective, it encompasses natural capital valuation principles as well as return on investment, property rights, tenure and other economic and legal factors.

To move forward with community development and increased standard of living for those communities, growth must be a factor, but that growth should be sustainable especially with reference to the valuation of the natural capital that can be affected by that growth. Indeed, it is even controversial as to whether and how standards of living should be increased. The Brundtland Commission[40] considered the obligations of developed countries to their developing counterparts, being concerned with global equity. Intergenerational equity and the future generations' inheritance of natural capital should also be considered.[41] Additionally, the concept that scientific resources could be combined with local indigenous knowledge to protect and restore natural capital resources should be considered.[42] For example, the private charitable organisation, Bush Heritage which owns and manages approximately 3% of Australia's wilderness areas, employs local indigenous people to train and assist their scientists in land management practices using traditional methods, such as regular cool burning.[43] Recently, an increased number of private and government organisations have started incorporating indigenous land management practices into the regular procedures.[44]

Key recommendations were made by Dr Finkel in the "Independent Review into the Future Security of the National Electricity Market" that energy governance should be focused on attracting "Increased investment in large, medium and small-scale variable renewable electricity (VRE) generation capacity and micro grids."[45] An example of this is small-scale solar panel infrastructure and battery storage. Dr Finkel's recommendation that the COAG should direct the Australian Energy Market Commission to undertake a review has been followed in part and much work has been and is continued to be done by regulatory bodies towards this goal.[46]

As COAG is actively working towards both the goal of Australian energy security through micro-grid infrastructure and the separate goal of improved infrastructure and economic participation by indigenous communities, it follows that those two goals can be complementary.

Modern infrastructure relies upon an electricity supply. Therefore the scope of this industry guide will initially focus on energy supply infrastructure. Smart city indicators are extensive and applied using different models. Some of the important indicators are pollution levels, innovation, CO_2 production, transparent governance, sustainable resource management, education facilities, health conditions, sustainable innovative and safe public transportation, pedestrian areas, green areas, waste facilities, household electricity supply, green fuels, political strategies and perspectives and availability of Information and Communications Technology (ICT) infrastructure.[47] Sometimes a smart city can be defined by different characteristics such as a smart economy, smart mobility, smart environment, smart people and smart governance.[48] Most studies and indicators are applied to urban areas, for example, the European Commission and European Union promote Intelligent Energy Europe in an attempt to boost clean and sustainable solutions in building city infrastructure.[49] By contrast, this industry guide focuses on remote communities and develops a set of remote community infrastructure indicators. These indicators are best practice decision-making tools that can be used to attract investment and guide and monitor investment decisions.

Property economics solutions to infrastructure supply are found in case studies. This industry guide provides a comprehensive qualitative analysis of the primary property economic solutions to infrastructure supply in remote regions.

The Sustainable Development Proposition

The SDP is a set of propositions or principles designed to analyse the sustainability of systems.

This SDP and the Sustainable Community Infrastructure Indicators (SCIIs™) analyse the attractiveness of an investment in renewable energy infrastructure for remote communities. The more attractive the investment, the greater the likelihood it will attract investment.

When attracting investment, the legal governance, community governance, social, environmental, technological and economic benefits should be analysed so as to demonstrate to investors the sustainability of the proposed investment. If the sustainability is assessed from a biased perspective, the investors will be less likely to fund the project. For example, if the investment is economically sustainable showing a healthy return on investment and short payback period but is not governed sustainably with an appropriate legal structure, the project is unlikely to attract investment. If the investment is not environmentally sustainable, community support might stall, and as demonstrated by the case studies, potential shareholders and investors are less likely to be attracted where there is no community support. If the technology used is unsustainable and cannot be maintained or upgraded within the system, the project will not sustain infrastructure in the long term.

The sustainability of a project requires analysis from an interdisciplinary skill set that most often requires analysis by a team of experts at the pre-development stage such as financial planners, accountants, valuers, lawyers, town planners, community leaders and engineers.

The interdisciplinary discourse provides a collective way of comprehending the world. It unveils the sought collectives or grids of intelligibility behind how people or institutions think and act.[50]

In order to create the SDP, the following areas are explored and evaluated in the next chapter:

1. Environmental sustainability:
   - Exploring the definitions of environmental sustainability using historical and doctrinal analysis, in particular, the precautionary principle and the principle of intergenerational equity;
   - How to value natural capital resources, both renewable and non-renewable;
   - Maintaining sustainable yields of natural resources;
   - Exploring a novel approach to the natural capital aggregate rule that can be applied to communities.
2. Sustainable governance:
   - A democratic justice system that affords liberty and human rights that support a sustainable community is explored;
   - Economic justice as it relates to energy poverty and infrastructure supply;
   - Legal and financial governance structures have been investigated and analysed in the case studies.
3. Sustainable technology:
   - A review of renewable energy technology available for non-network solutions has been conducted;
   - The relationship between sustainable technology, population growth, economic growth and an increased standard of living is explored.
4. Economic sustainability:
   - The macro concepts of economic growth as it relates to improvements in standards of living and an increase in population is analysed;
   - Economic sustainability from a traditional investment perspective has been analysed in the case studies and is addressed in the overarching recommendations;

- The case studies are used to evaluate systems that are economically self-sufficient and do not require ongoing public funding or charitable donations;
- The economic sustainability of renewable resources is addressed within the topic of environmental sustainability above.

A cost-benefit analysis of the case studies in conjunction with the evaluation of what makes up a Sustainable System has led to the development of a statement of propositional logic called the SDP. Propositional logic is the bridge between conceptual thought and mathematics. It is used for purposes such as computer programming and drafting legislation. The SDP consolidates the historical and doctrinal analysis and enables an interdisciplinary team to make decisions on infrastructure projects' sustainability pre-development phase or alternatively to monitor the sustainability of a project post-development phase. It ensures that the investment decisions are unbiased towards any one discipline within the team. The data fed into the Proposition is designed to be input by legal, financial, planning and engineering professionals as well as community members. A multi-disciplinary professional would be suitable to lead such a team. The SDP is made up of nine sub propositions. When those nine sub propositions are deemed sustainable in the long term then this indicates that the system is sustainable.

5. Sustainable system and Social benefits

A sustainable project, system or community is usually evidenced by social benefits, in this case, increased standards of living and reduced energy poverty.

This is demonstrated in the case studies where removal of energy poverty created an increased standards of living evidenced by improved access to education, employment, health services, safety and fresh food and medication. Additionally, community governance benefits were identified.

These benefits identified in the case studies were used to create the set of SCIIs™. An advantage of the SCIIs™ is that they are designed to be flexible and to work across a variety of community contexts. Guidance is to be provided in a transparent and understandable way such that local decision-makers can examine the trade-offs against the impacts across a decision-making matrix and decide for their own communities how to improve the investment proposals so as to maximise opportunities for a win-win outcome for their communities.[51] There are different tools that can be used to provide this sustainability analysis to communities. Leaders can implement or revisit sustainability assessments within their own policy and institutional contexts. This is in keeping with the underlying democratic principles of the western countries. It is this vein that the research into rural and indigenous communities has been conducted in this industry guide, to provide funding, investment and governance choices.

The SDP developed in this industry guide, working in conjunction with the SCIIs™, has been designed to allow for diversity across communities in rural and regional communities and as such is a flexible matrix. The by-product of this flexibility is that it can be applied across any number of regulated and unregulated markets and legal systems. This decision-making tool is not a "how to guide" for practitioners or experts but is a tool to form the basis of a more tailored plan for investment in a context-specific project. The choices made resulting from that analysis will depend on a variety of context-specific factors, such as the availability of resources, institutional capacities and legal structure choices made at governance levels in a variety of communities.

The end goal of the SDP is to develop economically diverse systems that are sustainable and attractive for investment. Those systems should enable the trajectory of increased standards of living and economic growth for communities while at the same time preserving the value of natural capital and cultural heritage for future generations.

What follows is a review of the nine propositions used to describe a Sustainable System (SS). See Table 1.1.

IF natural capital (NC) is enhanced, substituted, adapted OR remains constant AND intergenerational equity (IG) remains constant AND the precautionary principle (PP) is satisfied AND the democratic rule of law (DRL) is upheld AND human rights and liberties (HR) are upheld AND economic growth (EG) increases in line with the population (POP) AND technology (T) improves AND the standard of living (SL) increases in line with population (POP) growth THEN this is EQUAL to a sustainable system (SS).

$$((NCNC \vee NC) \wedge IG \wedge PP \wedge DRL \wedge HR \wedge (POP \leftrightarrow EG) \wedge T \wedge (POP \leftrightarrow SL)) \rightarrow SS$$

ABBREVIATIONS FOR TABLE 1.1:

1. *NCNC = Natural capital is enhanced, substituted, adapted OR remains constant*
2. *NC = Natural capital remains constant.*
3. *IG = Intergenerational equity remains constant.*
4. *PP = The precautionary principle is satisfied.*
5. *DRL = The democratic rule of law is upheld.*
6. *HR = Human rights and liberties are upheld.*
7. *EC = economic growth*
8. *POP = population growth*
9. *T = Technology improves.*
10. SL = The standard of living increases.

TABLE 1.1 The Nine Propositions in a Sustainable System to Create the Overarching Sustainable Development Proposition

The Propositions	*A Sustainable System*
FIRST proposition	*IF natural capital (NC) is enhanced, substituted, adapted OR remains constant* *(NCNC V NC)*
SECOND proposition	*AND intergenerational equity (IG) remains constant* *Λ IG*
THIRD proposition	*AND the precautionary principle (PP) is satisfied* *Λ PP*
FOURTH proposition	AND the democratic rule of law (DRL) is upheld Λ DRL
FIFTH proposition	AND human rights and liberties (HR) are upheld Λ HR
SIXTH proposition	AND economic growth (EG) increases in line with the population (POP) Λ (POP ↔ EG)
SEVENTH proposition	AND technology (T) improves Λ T
EIGHTH proposition	AND the standard of living (SL) increases in line with population (POP) growth Λ (POP ↔ SL)
NINTH proposition	THEN a Sustainable System (SS). → (SS))

Notes

1 Centre for Appropriate Technology Limited 2016, 8, 59.
2 Ibid.
3 Eras-Almeida 2019, 7139.
4 Guruswamy et al. 2016, Ch. 17.
5 Kanagawa et al. 2008, 2017.
6 Buergelt et al. 2017, 274–277.
7 Powercards are reported as being used in the Northern Territory, Wujal Wujal, Cape York, Hammond Island, Torres Strait and 32 other communities in conjunction with Ergon Energy Retail, from, Energy Queensland, "Energising Queensland communities" Annual Report 2017–2018, at p. 24; They are also reported as being used at the top end of Australia by Buergelt et al. (2017, 274–277).
8 Buergelt et al. 2017, 274–277.
9 Ibid.
10 International Energy Agency (IEA). Energy Access Outlook 2017: From Poverty to Prosperity; IEA: Paris, France, 2017. Feron, S.; Heinrichs, H.; Cordero, R.R. Are the rural electrification in the Ecuadorian Amazon sustainable? Sustainability 2016, 8, 443. From Eras-Almeida et al. (2019).
11 Sovacool, B.K. et al. New partnerships and business models for facilitating energy access. Energy Policy 2012, 47, 48–55, with reference to: Lemaire, X. Off-grid electrification with solar home systems: The experience of a fee-for-service concession in South Africa.

Energy Sustain. Dev. 2011, 15, 277–283. Urmee, T.; Harries, D.; Schlapfer, A. Issues related to rural electrification using renewable energy in developing countries of Asia and Pacific. Renew. Energy 2009, 34, 354–357.

12 Ibid.

13 The Organisation for Economic Co-operation and Development (OECD) "Guidance on Sustainability Impact Assessment".

14 Sovacool 2014. *Energy Research & Social Science*, 14.

15 Ibid.

16 Global Power System Transformation Consortium.

17 Reserve Bank of Australia 2020.

18 European Parliament 2019.

19 United Nations Environment Program (December 2008), 20.

20 Finkel, A.; Graves, M. *The Future of Renewables.*

21 Australian Government 2013. Australian Natural Resource Atlas.

22 The Chamber of Minerals and Energy of Western Australia 2021.

23 Ibid.

24 Australian Conservation Foundation 2008.

25 Fairley 2019.

26 Harvey 2020.

27 Townsend 2013, 178.

28 Ibid.

29 Ibid. 10.

30 Ibid.

31 Finkel 2017, 72.

32 Droege 2009, Ch. 2.

33 Ondraczek et al. 1130.

34 Closing the Gap Implementation Plan, 2021; See also Council of Australian Governments 2017. *National Indigenous Reform Agreement (Closing the Gap)*, 7.

35 Ibid.

36 Grant et al. 2018, 120; Martin 2016, 15.

37 Martin 2016, 15.

38 Grant et al. 2018, 120.

39 Cherbourg Aboriginal Shire Council 2021.

40 Report of the World Commission on Environment and Development: Our Common Future. Transmitted to the United Nations 1987.

41 Helm 2015, 37.

42 Martin 2016, 33.

43 Ibid. 202.

44 Queensland Government, Department of Environment and Science, Parks and Forests 2021.

See also, Robinson 2020.

45 Finkel 2017.

46 See, for example, ACCC, Restoring electricity affordability and Australia's competitive advantage: Retail Electricity Pricing Inquiry—Final Report, Canberra, June 2018.

47 Helm 2015, 37.

48 Ibid.

49 Ibid.

50 Sovacool 2014. *Energy Research & Social Science*, 13.

51 Ibid. 14.

2

SUSTAINABLE ENVIRONMENT

Introduction

The goal of the Sustainable Development Proposition is to develop economically diverse systems that are sustainable and attractive for investment. Those systems should enable the trajectory of increased standards of living that create social benefits and economic growth for communities while at the same time preserving the value of natural capital for future generations.

What follows in Chapters 2–5 is a literature review of the propositions used to describe a Sustainable System (SS) that make up the Sustainable Development Propositions.

This chapter explores environmental sustainability using historical and doctrinal analysis, the precautionary principle and the principle of intergenerational equity. Chapter 3 reviews sustainable governance, Chapter 4, sustainable economy and Chapter 5, sustainable technology.

This chapter analyses environmental sustainability, which includes, the first, second and third propositions in the Sustainable Development Proposition:

> IF natural capital (NC) is enhanced, substituted, adapted OR remains constant
>
> (NCNC V NC)
>
> AND intergenerational equity (IG) remains constant
>
> Λ IG
>
> AND the precautionary principle (PP) is satisfied
>
> Λ PP

DOI: 10.4324/9781003324669-2

FIGURE 2.1 A wind turbine near a body of water at sunset. Photo by Guillaume Meurice from Pexels.

Current state of the environment

Greenhouse gas emissions in the Earth's atmosphere have exceeded sustainable levels of 280 parts per million (ppm), and concentration reached 413 ppm in September 2021 but the seasonally adjusted mean is now 417 ppm.[1] However, not all governments agree on a safe level.[2] The target set by the Intergovernmental Panel on Climate Change (IPCC) states that direct CO_2 emissions from the energy supply sector are projected to double or triple by 2050 compared to the level of 14.4 $GtCO_2$/year in 2010, unless energy intensity improvements can be significantly accelerated. In their Fifth Report, the IPCC stated,

> In the last decade, the main contributors to emission growth were a growing energy demand and an increase of the share of coal in the global fuel mix. The availability of fossil fuels alone will not be sufficient to limit CO2eq concentration to levels such as 450 ppm, 550 ppm, or 650 ppm.[3]

The IPCC recognises that renewable electricity generation is a key component of cost-effective mitigation strategies in achieving low-stabilisation levels of CO_2 concentration and that focusing efforts on the electricity generation sector is a more rapid and efficient method than focusing on the industry, buildings and transport sectors.[4]

The IPCC's Sixth Report (the Sixth Report) was released on August 7, 2021, and for the first time addressed regional areas in particular.[5] This confirms the significance of this industry guide topic which is focused on regional and indigenous communities. The Sixth Report also addresses the role of technology and innovation for the first time.[6] This confirms the significance of including technology and innovation in the Sustainable Development Proposition developed in this industry guide, as will be explored further below. Additionally, the Sixth Report addresses the social aspects of mitigation for the first time.[7] This confirms the significance of the community and social aspects of sustainability addressed in the Sustainable Development Proposition and Sustainable Community Investment Indicators developed in this industry guide.

Other concerns in relation to fossil fuels are that they not only cause climate change but also lead to air pollution. The first attempt to reduce and regulate air pollution was the *Clean Air Act* (1965) (UK). Coal was initially used only by low socio-economic households in Britain, those who couldn't afford timber. As timber became scarce and expensive, Britain was forced to switch to a coal-fuelled economy.[8] Following a series of environmental and health disasters caused by coal mining and coal-fired pollution, Britain took its coal industry to the colonies, including Australia. Coal has been known and recognised by British *Clean Air Act* since 1965 as the cause of environmental harm and health hazard.

Electricity generation is the largest source of PM 2.5 particles which are hazardous to human health.[9] The Centre for Air Quality in Health Research Evaluation states that those living near lines and power stations are at most risk from air pollution.[10]

Coal-fired power stations are about 5% efficient. Modern gas-fired stations are slightly better at 40–20% efficiency. However, the world could be powered by renewable wind energy alone.[11] Coal and gas mining can cause irreparable damage to the underground water table, as demonstrated by the hydraulic modelling evidence before the Queensland Land Court in the New Acland Coal case.[12] Although the mining company tried to negotiate "make good" agreements with the local farmers and residents, their own witnesses admitted that a hundred years from now the mining company might not exist, let alone be able to pay for the cost of repairing damage to the water table, assuming technology could be invented in the future to repair such damage, which it was estimated was a remote possibility at best.[13] This case demonstrates the importance of considering all aspects of environmental pollution when designing energy infrastructure in rural areas.

It is unlikely that future regulators will be able to afford the costs of rectifying the damage caused by coal and gas mines in Australia, in particular the damage caused to ground water.[14] The cost should be borne by the polluting companies under a governing framework.[15] For example, RAJ-Stiftung, a German coal mining company is committing 300 million euros over the next few years to rectify pollution caused by its coal mining activities.[16] The clean-up requires pit water management, which is, transporting polluted water from the bottom of mine shafts into drainage stations. Additionally in regions that have subsided, the management of pooling polluted water must be conducted. Ground water must also be purified and monitored. The company has publicly committed to financing these clean-up operations on an ongoing basis.[17] This is an example of a company rectifying its environmental damage; however, others have declared insolvency after the mining stage is completed and before the clean-up stage commences.[18] An inquiry into the 2014 fire at Hazelwood found the rehabilitation cost of the mine to be $100 million dollars; however, the bond posted was only $15 million.[19] In many cases, the bonds are only posted for mines and not the power stations themselves. The mines and power stations should be regularly audited as to whether they are in the financial position to cover their decommissioning costs.[20] Additionally, the RAJ-Stiftung funding model could be deployed.

Furthermore, concentrated solar thermal energy with molten salt storage can replace up to 45% of existing gas infrastructure including gas used by the alumina industry. This can be done without any major changes to existing equipment or infrastructure.[21] The remaining gas infrastructure can be replaced with new large-scale solar and offshore wind installations.

Given the known environmental harms, and the serious health hazards of the coal industry, there is a strong argument to support the phase out of coal-fired power generation and the growth of renewable energy. The future of Australian economic independence could lie in the unlimited solar and wind resources available.

On the other hand, clean coal when combined with a responsible decommission and clean-up plan can provide affordable electricity contributing to a higher standard of living for communities that are grid connected. Many countries and especially regional towns rely upon coal to maintain their standard of living and will do so for the foreseeable future. Therefore, it is important that clean coal technology is supported and communities relying on coal are not disadvantaged or discriminated against as many of those communities are in developing towns with a low standard of living.[22]

In Australia, more than 75% of the existing coal-fired generation plants have passed their useful life.[23] Currently many coal-fired generation plants are being decommissioned and their supply deficits are being met by solar and wind generation farms as announced by AGL®, Origin Energy® and GDF Suez®.[24] This decommissioning process must be managed in a responsible and orderly way. As outlined by former Chief Scientist, Dr Finkel, this process should include, "A long-term emissions reduction trajectory for the electricity sector. An obligation for all large generators to provide at least three years' notice of closure. A credible and enduring emissions reduction mechanism."[25] Australia will also need adequate renewable energy generators and transmission grids to replace the coal-fired generation plants before they are decommissioned.[26] If Australia decommissions outside of these recommendations, it could cause significant electricity supply shortages and damage to the economies of remote towns as well as decreased standards of living and energy poverty among the least fortunate.

The European Union (EU) often tries to interfere with the decarbonisation policies of countries outside of the EU. Decommissioning the coal industry in Australia should be a matter for Australia as region-specific issues need to be addressed in a responsible way during the transition period to ensure remote communities are not plunged into energy poverty and the resultant reduced standards of living. Likewise, developing nations should be supported to design their own energy transition structures that suit their community needs.

Australia has a track record of meeting its international treaty obligations in relation to greenhouse gas emissions reduction targets, exceeding its obligations under the Kyoto Protocol (an international agreement under the United Nations Framework Convention on Climate Change)[27] for the first commitment period from 2008 to 2012.[28] Australia ratified the subsequent Paris Agreement on November 10, 2016. Australia has committed to reducing national emissions by 26–28% on 2005 levels by 2030 under the Agreement.[29] The Paris Agreement was informed by the Fifth Scientific Assessment Report from the United Nations body, the IPCC.[30] An interest in renewable energy in infrastructure has increased following the further Special Report published by the IPCC in 2018 (the Report).[31]

Sustainable infrastructure that supports economic growth and an increased standard of living, thereby creating social benefits while meeting the obligations of the *Paris Agreement*, will be an important aspect of future development and the supply of infrastructure to remote communities. More specifically, the Report highlights that community approaches, informed by indigenous knowledge in relation to sustainable development, can accelerate widespread behaviour change and limit global warming to 1.5°C. However, "public, financial, institutional and innovation capabilities currently fall short of implementing far-reaching measures at scale in all countries."[32] The Report notes that "lifestyle and behavioural change and innovative financing mechanisms can help their mainstreaming within sustainable development practices. Preventing maladaptation, drawing on bottom-up approaches and using indigenous knowledge would effectively engage and protect vulnerable people and communities."[33]

The Report recognises that increasing health, alleviating poverty, improving agriculture and advancing gender equality will cause a trade-off in energy consumption, agriculture and clean water. Consequently the Report recommends sustainable development as having the potential to "significantly reduce systemic vulnerability, enhance adaptive capacity, and promote livelihood security for poor and disadvantaged populations."[34] Because,

> ... some of the worst impacts on sustainable development are expected to be felt among agricultural and coastal dependent livelihoods, indigenous people, children and the elderly, poor labourers, poor urban dwellers in African cities, and people and ecosystems in the Arctic and Small Island Developing States.[35]

The Report notes that in expanding cities where poverty is high, there is an "opportunity to benefit from recent price changes in renewable energy technologies to enable clean energy access to citizens .... This will require strengthened energy governance in these countries."[36]

As humanity is increasingly questioning the availability of resources and the impacts of climate change, economists are tempted with the zero growth economy models.[37] A zero growth economy could result in consequences of a radical nature such as decreased standards of living, food production problems, rationing of resources and limits on human rights and freedoms.[38] This is particularly problematic for communities that already have a marginal standard of living. According to Helm, the zero growth economy model

> like most if not all utopias, it suffers from two serious defects: it is not necessarily desirable; and it is never going to happen. It is at best a distraction, and at worst an excuse for not engaging with the world as it is, rather than the world as its advocates would like it to be, with all the coercive restrictions on freedom, enterprise and choice that it entails.[39]

The development of energy infrastructure from a property economics approach can provide alternative models to the zero growth model. As outlined in the Report, new financial mechanisms backed by responsible governance can take advantage of the decreased cost of sustainable infrastructure supply to vulnerable communities and provide models for economic growth that are aimed at increasing standards of living and improving democratic freedoms.

Internationally and domestically, there are tens of thousands of community-based sustainable energy infrastructure projects. In many case studies, the renewable resources are contributing significantly to regional economies. In other case studies,

FIGURE 2.2 White wind turbines on a hilly farm. Photo by Pixabay from Pexels.

although the infrastructure is environmentally sustainable, the projects are economically, socially or systemically unsustainable.[40] For infrastructure to be sustainable, it must also have Sustainable Systems, that is, legal, governance and financial systems. The technology must also be sustainable, in that it must be capable of being upgraded and maintained and it must be governed in a socially sustainable way.

The first consideration for sustainable infrastructure reviewed below is environmental sustainability. The analysis of environmental sustainability is the subject of much debate. Consequently, we begin the discussion in the next section by reviewing the term "environmental sustainability."

Environmental sustainability defined

The precautionary principle

It helps to have a definition of environmental sustainability, and if you turn to the law, the primary legal principle that protects natural capital is the precautionary principle. It is applied in project planning and embodies the adages "Stop and think before you act" and "look before you leap." Stopping a project due to the precautionary principle does not mean that the particular technology or process will be halted for all time; it means that it will be halted or delayed until such time as sufficient evidence, experimentation or improvement are made to satisfy the principle. For example, Maori leaders stopped scientists from the US from experimenting with gene editing technology on ticks in New Zealand as it is too risky from a precautionary principle standpoint in terms of risks that eugenics could have on a possible subsequent ecological collapse. Three primary concerns were expressed, firstly, that international eugenics companies were assuming that they could experiment in indigenous communities and there was a feeling that there was an attempt to take advantage of the Maori people. Secondly, eugenics was a concern for the Maori elders because of its historical links to racist experimentation. Thirdly, those elders expressed a traditional belief that all living things were connected, and by altering artificially the genes of one creature in the ecosystem, the whole balance of the land would be undermined and this could cause a collapse of a system.[41] This last point is a traditional belief that is similar to the precautionary principle. It is important to understand that the precautionary principle is not only applied by scientists, lawyers and governments but also by community members and their leaders, who are often aware of local risks that scientists and experts from outside the community have not considered.

In Australian Indigenous culture, the concept of "country" demonstrates a connection to the land and the environment.[42] Country is part of the "Dreaming" and holds the law and knowledge of the people.[43] Everything within country is living. There is no division between animate and inanimate.[44] The Aboriginal concept of country is all-encompassing. It includes the land, the flora, the fauna, the people, the culture, the water, the sky and the spirits.[45] Country is the interconnected web of life that is integrally tied to ancestral lands.[46] Connectedness to country is the foundation of the relationship with kin. Aboriginal people believe that country "owns" us and that we have obligations to care for country.[47] As country incorporates the natural environment as well as the human element, it links the social, community and spiritual values to the environment. This means that the spiritual aspects, value systems and customary laws are all part of country. This can be contrasted with the western view of nature which often views community, spirit and culture as separate from the environment.[48] However, traditional Christian values often refer to humanity as being custodians of the plants and animals in service to God.[49] Similarly, Indigenous Australians view themselves as custodians of country with obligations to country, including the natural living, spiritual and non-living aspects of the environment.[50] An example of this obligation can be seen in the origin dreaming story of the first Worimi people who began life as dolphins, Guparr and Wubarray, who created hiding places for the fish to protect them from being depleted. These dolphins were rewarded by the creation spirit, Baiame, who gave the dolphins the ability to walk on land and transformed them into the first humans.[51]

Traditional Indigenous sustainable practices on country have been incorporated into the Australian farming industry, for example, Indigenous and modern fishing law prohibit killing female crabs and crayfish and require small fish to be released.[52] Fishing practices such as fish traps are an example of "economic and social organisation" that aims to "sustain the fishery."[53] The fish traps systems at Brewarrina, northwest of Sydney, are attributed to the creative spirit Baiame. The people arranged rocks into patterns that trapped the fish so they could cooperatively harvest fish with spears. The system was built using arches and keystones to strengthen the fish traps. It is unknown how old the fish traps are. Estimates range between 3,000 and 40,000 years. Passages in the fish traps allowed upstream fisheries from different families to also benefit from the breeding stock.[54]

The concept of country is therefore similar to the concept of natural capital. And sustaining natural capital is similar to the concept of caring for country. The Australian Indigenous perspective of "natural capital" means "valuing the earth and the raw materials it provides for as an essential part of conservative economics."[55]

The precautionary principle is applied to renewable natural capital. This is because if the stock of a renewable resource is depleted beyond the threshold for which it can self-replicate or renew, there is no way back without very large asymmetric costs, if there is a way back at all.

The precautionary principle states: "… Where there are threats of serious or irreversible environmental damage, lack of full scientific certainty should not be used as a reason for postponing measures to prevent environmental degradation."[56] Therefore, the principle will be triggered at law in Australia where two conditions are present: firstly, the threat of serious or irreversible environmental damage, and, secondly, a lack of scientific certainty as to that damage. When applied and triggered, the precautionary principle states that public and private decisions should be guided by:

- Careful evaluation to avoid, wherever practicable, serious or irreversible damage to the environment; and
- An assessment of the risk-weighted consequences of various options.[57]

In practice, this principle is complicated and dependent on the facts of each project individually. The decision to take precaution can hang on analysis of whether a threat is theoretical and unlikely or actual serious or irreversible.[58] Expert witnesses are involved when the precautionary principle is disputed. Strong disagreement among experts is usually a good reason as to why court would apply precautionary approach.[59] It is worthwhile noting that the precautionary principle has not been used in Australia to prevent the development of infrastructure due to air emissions exacerbating the greenhouse effect.[60] However, avoiding or lessening climate change has been included as an object in most planning instruments.[61] The precautionary principle has been applied, however, to protect water or endangered species.[62] It has also been applied in a decision against allowing the continued operation of a colliery. In *SHCAG Pty Ltd v Minister for planning and infrastructure [2013] NSWLEC 1032*, the court said at [89 to 90]:

> we are satisfied that the precautionary principle is activated as the risk of significant environmental harm currently remains uncertain, based on the evidence before us, as the proposal may result in the following:
>
> 1. the dewatering of the Hawkesbury sandstone ground aquifer, which would change its ecology and may prevent future access to water for irrigation purposes; and/or
> 2. An adverse impact on the health of the Wingecarribee River by discharging pollutants in the water discharged from the mine.

Environmental law is complex, especially when it intersects with infrastructure and mining. For example, the objects of the Environmental Protection Act (1994) (EPA) are to "protect Queensland's environment while allowing for development that improves the total quality of life, both now and in the future, in a way that maintains the ecological processes on which life depends (ecologically sustainable development)."[63] The *Planning Act (Qld) (2016)*, s 5, lists among multiple sustainable development objectives and ways to achieve those objectives, among them, the precautionary principle, climate change avoidance and intergenerational equity. There are similar laws found in most western planning jurisdictions worldwide. Similarly, the principle of intergenerational equity is also found in many planning regulations.

Intergenerational equity

Intergenerational equity has been described in economic terms as, "using the interest produced by our assets while keeping the capital intact."[64] One of the ways in which communities can manage capital resources is through sustainable development practices.

The concept of sustainable development has formed within most disciplines. The Brundtland Report's definition of sustainable development has become the most widely accepted definition, namely,

> Sustainable development is development that meets the needs of the present without compromising the ability of future generations to meet their own needs: the concept of 'needs', in particular the essential needs of the world's poor, to which overriding priority should be given; and the idea of limitations imposed by the state of technology and social organisation on the environment's ability to meet present and future needs.[65]

Intergenerational equity is a concept familiar to most cultures. Native Americans call this principle, "The Great Law of Peace of the Haudenosaunee", which states that chiefs consider the impacts of their decisions on the seventh generation.[66]

Indigenous Australians describe intergenerational equity as

> every product we use be stamped with our determination that our great-grandchildren can enjoy them in the future. This means our care must be extended to soil, water, food and the products we have created from the resources of the earth.[67]

These practices are examples of "caring for country."[68] In Haasts Bluff, the Northern Territory local people harvest bush tucker as a customary activity. The customary rules are that they "never take more of a resource than they and their families need."[69] Thus, they are maintaining natural capital for future generations.

The principle of intergenerational equity provides that "current generations hold the environment in trust for the benefit of future generations" or that 'the present generation should ensure that the health, diversity and productivity of the environment is maintained or enhanced for the benefit of future generations'."[70]

We can describe that wealth as being natural capital assets, such as clean air, water, food sources and access to biological resources necessary for our survival as a species. We are all the present generation and collectively need to take responsibility for this principle. For example, in the Philippines, plaintiffs were held to represent the interests of future unborn citizens.[71] It has also been considered in Australia in the preservation of Aboriginal cultural heritage.[72]

More detail on the recent history of the application of intergenerational equity in resource law can be found in both the Acland Coal and Adani mining cases,

> intergenerational equity involves considerations of equity within the present generation, such as uses of natural resources by one nation-state (or sectors or classes within a nation-state) needing to take account of the needs of other nation-states (or sectors within a nation-state).[73]

Economists also call it Eco-equity:

> equity between peoples and generations and, in particular, the equal rights of all peoples to environmental resources. At the heart of nearly all sustainability goals is the belief that there should be a fair distribution of resources both

FIGURE 2.3 A solar farm. Photo by Red Zeppelin from Pexels.

within and across generations. Eco-equity focuses on our social responsibility for the future generations who will bear the consequences of excessive consumption of scarce resources and environmental degradation.[74]

Valuing natural capital

Environmental sustainability also pertains to the sustainability of natural resources to ensure sustainable yields, as explored and evaluated below. To do this, the concepts of renewable and non-renewable capital must be explored.

In GDP (gross domestic product) terms, developed countries have experienced continuous rapid growth in standards of living over recent generations. This does not account for damage that might occur to renewable natural capital availability for future generations. For example, the availability of clean water and clean air for future generations should be factored into the predictions for standards of living for future generations if we continue on the current trajectory. We should not leave future generations with pollution, waste, smog, nuclear waste and dead ecosystems.[75] GDP ignores the natural resource asset base and also ignores the liabilities.

Debt-adjusted GDP is not accounted for, that is, the debt the future generations will be responsible for is not counted.[76] To know whether we are better off than last year, assets, liabilities and returns on assets should be taken into account. "If you borrow too much from the environment you accumulate an ecological debt that needs to be paid down."[77] National income accounts do not measure the state of the assets, the state of the natural capital assets, renewable and non-renewable. For example, the exploitation of non-renewable natural capital could result in investment in economic rents in other renewable capital assets. This would mean that future generations who have forgone the opportunity to capitalise on the non-renewables will be compensated by the renewable capital assets.[78]

One of the considerations for sustainable development is the management and valuation of the sustainable yields or resource sustainability. This requires valuation methodology, as discussed below.

Firstly, renewable and non-renewable natural capital must be distinguished. Non-renewable natural capital refers to finite resources such as oil and gas. Renewable natural capital refers to the free, self-replicating resources, such as fish, trees, sunlight and air. Renewable resources, although self-replicating, might in fact disappear faster than non-renewables, if they are mismanaged.

There are many renewable resources that nature yields to us that are being depleted beyond a point of sustainability. Resources such as fish and trees are renewable but are mismanaged in some parts of the world. We can catch and eat fish, and the fish will keep repopulating. Trees can be cut down and regrown. The capital continues to be renewable so long as we don't over fish or drive the resource beyond the margin of sustainability to a threshold where the fish or trees can no longer replicate themselves. That would be a case of depletion of a renewable resource.[79]

Sustainability is less concerned with the depletion of non-renewable natural capital and more concerned with the depletion of renewable natural capital that should be free and self-replicating. For example, concerns about the depletion of oil reserves are less concerning economically than the cost of air pollution, mitigating global warming or the cost of famines and natural disasters. Concerns about the depletion of oil reserves are limited to concerns about who gets the economic benefits of the finite resources remaining for exploitation.[80] By contrast, concerns about the cost of air pollution and global warming impact all of humanity. According to the Centre for Air Quality and Health Research and Evaluation, more than 3,000 people die from air pollution in Australia every year.[81]

More examples of renewable natural capital are clean air, the ozone layer, edible plants and animals, rainforest ecosystems, biodiversity, sunlight and clean water. All of this free, self-replicating natural capital benefits the survival of our species and is vulnerable to depletion beyond sustainable levels if overexploited. At an extreme level, when natural capital cannot self-replicate, historically mass extinction events have occurred.[82]

One method of valuing natural capital is through diminishing marginal returns. For example, if the Easter Islander's had recognised, as they deforested their island, the diminishing marginal returns on the renewable forest asset, this could have been an economic indication to halt logging to allow regeneration before the natural capital collapsed beyond the threshold of self-replication.[83]

Sustainable economy is not about how much renewable natural capital you have, it is about how much you can keep, how hard it works for you (efficiency can be increased with technology) and how many generations you keep it for (if it is renewable and well-managed, the generations you keep it for should be unlimited). Zero growth is not necessary for a Sustainable System. A preferable goal is sustainable growth, namely, growth that supports renewable natural capital at a self-replicating or self-sustaining level. This principle must be flexible, as nature evolves and species adapt to environmental change. Supporting natural capital in a static unchanging state is not realistic. Adaptation within ecosystems must be accounted for.

Traditionally, cultures have encouraged the cultivation of nature as an important part of survival. Much of our progress is embodied in this activity. Human interventions mean that ecosystems adapt and change. New ecosystems emerge as humans build habitats. For example, artificial reef ecosystems have formed at the base of many offshore wind farms and scavenger populations have increased near roads, due to roadkill abundance.[84]

The basis of natural capital valuation is in science, namely, the following ecological principles:

1. Energy can neither be created nor destroyed. When energy is transferred from one form to another, it degrades, releases waste and dissipates heat.
2. Natural systems are interdependent. Each action within a system has an impact on the functioning of the system as a whole.
3. Systems that are tightly interdependent lack resilience (monoculture). Diverse systems are more resilient (permaculture).
4. All species, including humans, are subject to the laws of nature.[85]

Natural capital assets balance sheets

In practice, maintaining a balance sheet of all natural capital assets is enormous, although the Federal Government is moving toward this practice in Australia. Assets such as water are valued by the Australian Government Department of Agriculture, Water and the Environment. The same can be done for renewable energy assets using sources such as the Renewable Energy Assets Atlas of Australia.[86]

The Australian Bureau of Statistics has begun measuring "environmental assets": "The present value for environmental assets has increased by 1% to $6,563b at current prices[87]:

- Land increased by 1% to $5,921b
- Minerals decreased by 2% to $386b
- Energy increased by 6% to $244b
- Timber increased 2% to $12b."[88]

By comparison, the UK Office for National Statistics (ONS) has estimated the value of natural capital at 1 trillion pounds. Renewable energy generation grew from 5% of all electricity generation in 2008 to 35% in 2018, and in 2018, coal production was 16 times less than in 1998.[89]

To reiterate the above, in order to meet the natural capital aggregate rule for a community or nation, maintenance of the value of the sovereign natural capital or increase in the value of the sovereign natural capital must occur.

This is impossible to achieve when a nation's assets are valued using the traditional GDP method.

Alternatively, it is proposed that the following innovative formula be used: an adjustment of the non-renewable assets, net of the costs of extraction, could be subtracted from the GDP. This should be compensated for by equal investment in renewable natural capital to meet the natural capital aggregate rule, the goal being, maintenance of the value of the sovereign natural capital or increase in the value of the sovereign natural capital.

This process encourages innovation, investment and sustainability and is a preferable policy solution to a taxation or penalty-based system such as a carbon tax.

The step-by-step process would be as follows:

Step 1. Non-renewable assets depleted annually are valued and then adjusted, NET of the costs of extraction = Adjusted Non-Renewable Asset Value (ANNAV).

Step 2. Adjusted non-renewable asset value is subtracted from the GDP = True Value of the GDP taking into account the non-renewable asset base that is depleted annually (TVG).

Step 3: Investment in Renewable Natural Capital is valued annually (IRNC)

If the IRNC is equal to or greater than the ANNAV then the total natural capital aggregate rule has been maintained or improved even where non-renewable assets have been depleted.

Step 4: Value the IRNC and adjust the GDP by adding the IRNC to the GDP = True Value of the GDP taking into account the renewable asset base that is invested in annually.

$$\therefore \text{GDP} - \text{ANNAV} + \text{IRNC} = \text{True value GDP}$$

If the IRNC is equal to or higher than the ANNAV then the natural capital balances are meeting the aggregate natural capital rule.

$$(IRNC \geq ANNAV) \exists\ NATURAL\ CAPITAL\ BALANCES\ meet\ AGGREGATE\ NATURAL\ CAPITAL\ RULE$$

This methodology could be used by governments to measure their contributions to meeting obligations under the *Paris Agreement*. This method might encourage investment in sustainable industry rather than just penalising polluters (as per a carbon tax system).

The aggregate natural capital rule is defined as follows:

- *Weak aggregate natural capital rule*: the aggregate level of renewable natural capital should be kept at least constant and there should be general capital compensation for the depletion of non-renewables.
- *Strong aggregate natural capital rule*: the aggregate level of renewable natural capital should be kept at least constant and the value of the economic rents from the depletion of non-renewable natural capital should be invested in renewable natural capital.[90]

The Brundtland definition of sustainability covers all assets not only natural capital. Infrastructure is another asset class. Energy infrastructure, for example, is essential to the functioning of an economy. It is necessary for the provision of primary goods and is a system not suitable for marginal analysis in that it cannot be removed from society if the costs outweigh the benefits. Electricity is either provided or not. Both electricity and natural capital are necessary for the modern economy.

> ... the key point is that renewable natural capital comes in systems too, and it can usefully be thought of as part of the core infrastructure. It is a special type of asset analogous to the special role of, say, electricity systems.

Professor Helm, fellow in economics at Oxford University, explains

> The modern economy needs to rely upon a continuous, high-quality and reliable supply of electricity. Natural capital is a similarly necessary input to a well-functioning economy. Without electricity it would be hard to imagine how the coming economic growth in this century could materialise. Similarly, a hotter climate with half the biodiversity may not be good news for economic growth and all those extra billions of people. The aggregate capital assets in these man-made infrastructures also need protecting and enhancing.

Combining the sustainable growth of both assets, *aggregate capital infrastructure assets* and *renewable natural capital assets* would make for a sustainable infrastructure project.

In the global economy, balance needs to be struck between specialisation and diversity. When a country specialises in agriculture or services and technology, it causes reliance on imports from other countries to support that specialised industry. For example, an economy that is based on services and technology with no significant agriculture relies heavily on other countries for imports of primary resources (food, raw materials and water).[91] The trend of globalisation has occurred because it is economically efficient for consumers.[92] However, it is not economically efficient in relation to the management of natural capital as it leads to the depletion and degradation of the resources in one country (commonly developing countries) for the benefit of other countries who import those natural capital resources.[93] This is why when exporting natural capital, it is important to keep accurate balance sheets of the national natural capital using the aggregate natural capital rule.[94] When this practice is adopted, a nation should be able to embrace global trade of natural capital and maintain the natural capital assets for future generations, provided country risks are evaluated and mitigated.

Valuation methodology: natural capital assets

There are various accepted valuation methods used to value property, infrastructure and natural resources. These are important to ensure the economic sustainability of a project.

Present value (discounting)

The net present value method is useful for valuing natural capital as it applies to commercial property ownership. This method is accurate in determining, for example, whether the loans needed to purchase a commercial property are serviceable. From an intergenerational equity perspective if natural capital is valued at its net present value, it would demonstrate the natural aggregate rule of value as it needs to be maintained in order to make the natural capital "debt" to future generations "serviceable."

For example, if the value of a crop of wheat is $500 per hectare, the cost of labour is $300, the cost of goods and seed is $50, the profit is $150. If the interest rate for cash saved in the bank is 2% and shares return an average of 5%, the riskier farming venture (exposed to weather, drought and flood events) might need a 10% profit margin to attract investment. Given these factors, the capital value per annum is discounted at 10%.[95]

However, other natural capital costs can be considered: the cost of restoring the farmland in relation to any damage pesticides cause to nearby water systems or the cost of restoring the soil so that the land can continue to be farmed for many generations. When the cost of damage to renewable natural capital is factored in and discounted, the economic incentive to farm using organic or permaculture techniques might outweigh the upfront cost of building the infrastructure that supports those systems,[96] especially considering the diminishing returns on yield of production in agriculture intensification and the increasing returns on yield of production with permaculture.[97]

The Nestlé Waters group which owns Perrier and Vittel is the largest bottled water company in the world by revenue. By the 1980s, the source area feeding the springs for their bottled water was becoming polluted.[98]

The company negotiated 30-year contracts with the farmers who agreed to use less intensive farming methods reducing herbicides and other pollutants. Each farmer was paid approximately $230 per hectare for seven years adding up to about $155,000 for the average farm. The company provided technical assistance, farm equipment and construction. Organic farms were also supported over approximately

500 ha of vineyards and wheat fields. This case study demonstrates Coase Theorem which is an alternative to government regulation of negative externalities such as a carbon tax. The theory suggests that where private property rights exist parties will bargain and create private transactions that not only benefit their own property rights, but also provide a public good, such as in this case a clean watershed.[99]

This Coasean bargain employed by Perrier to clean the pollution to water caused by farms demonstrates the cost to be discounted for environmental damage. Although in that scenario the farmers benefited financially because Perrier paid the cost of conversion to organic farming plus a bonus to each farmer who continued the practice.[100] The company bore the cost because they benefited financially from the clean water. A similar Coasean bargain could be struck between the tourism and hotel industry in the Great Barrier Reef and the local sugar cane farms. If the local hotel industry paid the farmers to switch to organic farming, both parties would benefit. Property rights approaches have also been used successfully to manage fishing resources in countries such as Alaska, New Zealand and some northern European countries.[101]

Present value was first formulated by Fibonacci. Irving Fisher applied it to investment management of shareholder interests in the early 1900s.[102] The present value method is also called discounting. This introduces a time dimension when thinking about efficiency and so is useful in dynamic industries, and environmental policy settings. This is because an efficient policy maximises the present value of net benefits to society. It can be a yardstick to measure benefits and costs occurring at different points in time.[103] The present-day environmental policy has to take into account all of the benefits and costs no matter when they might occur in the future. This is consistent with the value of the dollar, which will not be the same value as a dollar received at a different point in time in the future. This is what the time value of money takes into account.[104] The costs and benefits expected to occur in the future are discounted. This gives us the present value. The choice of a discount rate to be applied is complicated and requires expert valuation, cross-checking and acceptance of the wide range of views on the discount rate to be applied depending on the public policy, or investment being made.[105] The choice of discount rate can significantly change the present value of an asset. Therefore, this should not be conducted lightly in relation to public policy.

Cost-benefit analysis

Present value method can be cross-checked using a standard cost-benefit analysis approach. The cost of permaculture/organic farming is higher than pesticide use. However, the ongoing benefits long term of preserving the natural capital of the soil and water so that it doesn't need to be restored might outweigh those initial capital costs. The cost-benefit analysis was used by the US Environmental Protection Authority to decide whether to reduce lead in fuel.[106] It was not only the primary costs of phasing out leaded fuel and installing new equipment at refineries that was taken into account. But also the other benefits of phasing out lead. Firstly, reduction in damage to cognitive and physiological development in children. Secondly, reduction in exacerbation of high blood pressure in men. Thirdly, reduction in emissions and pollution. Finally, lower costs of engine maintenance and improved fuel economy. This cost-benefit analysis succeeded in supporting better regulations for lead-based fuels.[107] This is one example of how a cost-benefit analysis can be a useful framework for initial discussions on policy change. It should be noted that this method only succeeds if the participants disclose all of the costs, including damage to natural capital and public goods. This method is used in the case studies in a subsequent chapter.

Return on investment (ROI)

Return on investment (ROI) is another method that can be used. The ROI for solar PV is much greater than for a diesel generator in off-grid communities, as shown by the case studies.

Return on investment is the method used when assessing a traditional energy project such as a power plant. The initial upfront cost of installing the plant and ongoing costs against the profit generated over the lifetime of the plant gives the return on investment rate. This rate is not discounted for depletion of the State's natural capital assets. The cost of decommissioning and clean-up of waste is also historically underestimated. For example, an inquiry into the 2014 fire at Hazelwood found the rehabilitation cost of the mine to be $100 million dollars; however, the bond posted was only $15 million.[108]

Renewable energy is attractive for investment when an investment is analysed using the return on investment model, coupled with the valuation of natural capital saved (employing the present value method to value the clean air, water and the greenhouse gas emissions saving).

Return on investment is a good method for attracting investment. Loans are more readily obtained if an ROI can be demonstrated. Other methods are not as useful for obtaining business loans in all cases.[109]

In summary, all economic factors required to be considered should be taken into account. Additionally, cross-checking various valuation methodologies is good practice.

Therefore, to assess the economic viability of a renewable energy project, the primary valuation methods to be cross-checked against one another are present value, cost-benefit analysis, return on investment and the environmental sustainability of the natural capital assets.

Through investment in environmentally sustainable practices that are in line with the precautionary principle and the principle of intergenerational equity, a sustainable project can be managed. This will attract investment in technology and infrastructure that preserves natural capital.

Summary

In summary, the Sustainable Development Proposition aims to achieve goodwill and equity between generations. Respect between generations so that they can

FIGURE 2.4 A modern cow barn with solar panels on the roof. Photo by Wijs (Wise) from Pexels.

learn from and help each other is a hallmark of a long-term sustainable community. To this end, the precautionary principle and intergenerational equity should be taken into account when making decisions that impact the environment. Further, the maintenance, substitution or enhancement of the natural capital (for example, through evolving natural capital or replacement of extracted natural capital) should be considered to ensure that natural capital is available for the benefit of future generations. Valuation of natural capital can be useful in this regard.

Thus, natural capital should be enhanced, substituted, adapted or remain constant, intergenerational equity should at the minimum, remain constant and the precautionary principle should be considered when developing communities.

IF natural capital (NC) is enhanced, substituted, adapted OR remains constant

(NCNC V NC)

AND intergenerational equity (IG) remains constant

Λ IG

AND the precautionary principle (PP) is satisfied

Λ PP

Notes

1 US Department of Commerce 2019.
2 Ibid.
3 United Nations Intergovernmental Panel on Climate Change (IPCC). *Fifth Assessment Report*, Summary for Policy makers, 2019.
4 Ibid. 21.
5 United Nations Intergovernmental Panel on Climate Change (IPCC). *Sixth Assessment Report*, 2021.
6 Ibid., see also Fact Sheet at www.ipcc.ch/site/assets/uploads/2021/06/Fact_sheet_AR6.pdf
7 Ibid.
8 Hallett et al. 2011, 78.
9 National Pollutant Inventory Emission data, 2015.
10 Centre for Air Quality and Health Research and Evaluation 2013, 4.
11 Girardet 2008, 181.
12 *New Acland Coal Pty Ltd v Ashman & Ors and Chief Executive, Department of Environment and Heritage Protection* (No. 4) [2017] QLC 24.
13 Ibid.
14 Walters 2016.
15 Wynn and Julve 2016.
16 RAG-Stiftung, Annual Report 2018.
17 Ibid.
18 Denniss and Campbell 2015.
19 Ibid.
20 Ibid.
21 Australian Renewable Energy Agency, "Integrating concentrating solar thermal energy".
22 International Energy Agency, "Clean Coal Technologies".
23 Nelson et al. 2014, 15.
24 Chambers 2015; Parkinson 2015; Edis 2015.
25 Finkel 2017, Chs. 3 and 7.
26 Ibid.
27 Australian Government, Department of the Environment and Energy 2015.
28 United Nations Climate Change, *The Kyoto Protocol* (signed by Australia in 1998, ratified in 2007), made under the United Nations Framework Convention on Climate Change (1994).
29 The Paris Climate Agreement (2015). *United Nations Convention on Climate Change* (1994) (UNFCCC).
30 United Nations. *Fifth Assessment Report* (2015)
31 The United Nations. *Special Report* (2018).
32 Ibid. 22–23, 38.
33 Ibid. 40.
34 Ibid. 44.
35 Ibid. 43.
36 The United Nations, *Special Report* (2018), citing (Westphal et al., 2017; Satterthwaite et al., 2018); (Cartwright, 2015; Watkins, 2015; Lwasa, 2017; Kennedy et al., 2018; Teferi and Newman, 2018); (Eberhard et al., 2017) at 331.
37 Barrett 2018.
38 Helm 2015, 37.
39 Ibid. 37.

40 Hicks and Ison 2018, citing (Allan, 2014; Entwistle et al., 2014; Lantz and Tegen, 2009) and to transitions away from fossil fuels and nuclear energy (Bauwens, 2016; Boon and Dieperink, 2014; Olesen et al., 2002; Schreuer and Weismeier-Sammer, 2010; Toke et al., 2008).
41 Egender and Kaufman 2019.
42 Sveiby 2009, 341–356.
43 Gammage et al. 2021, 11.
44 Ibid. 11.
45 Kennedy et al. 2016, 17–18.
46 Ibid.
47 Ibid.
48 Ibid.
49 Genesis 1:26; Psalm 24:1.
50 Kennedy et al. 2016, 17–18.
51 Russell 2021.
52 Gammage et al. 2021, 11.
53 Pascoe 2014, 72.
54 Ibid. 73.
55 Gammage et al. 2021, 102.
56 Bates et al. 2019, 8.56.
57 See, for example, The *Planning Act* (Qld) (2016), s 5; *Telstra Corporation Ltd v Hornsby Shire Council* [2006] NSWLEC 133, *Paltridge v District Council of Grant* [2011] SAERDC 23, *Elliott v Brisbane City Council* (2002) QPELR 425; *St Helens Area Landcare and Coastcare Group Inc v Break O'Day Council* [2007] TASSC 15, From Bates, G. M. (Gerard Maxwell) et al. (2019) *Environmental law in Australia*. 10th edition. Chatswood, NSW: LexisNexis Butterworths at 8.56.
58 *Telstra Corporation Ltd v Hornsby Shire Council* [2006] NSWLEC 133.
59 *Hancock Coal Pty Ltd v Kelly & Ors* [2013] QLC 9.
60 *Greenpeace Australia Ltd v Redbank Power Co Pty Ltd* [1994] NSWLEC 178.
61 See, for example, The *Planning Act* (Qld) (2016), s 5.
62 *Telstra Corporation Limited v Pine Rivers Shire Council & Ors* [2001] QPE 014, *De Lacey & Anor v Kagara Pty Ltd* [2009] QLC 77.
63 See application in *Donovan v Struber & Ors* [2011] QLC 45 [13–16].
64 Bates, *Environmental Law in Australia*, LexisNexis, 9th ed (2016) at page 317. Bates also refers the reader to E. Brown Weiss, *In Fairness to Future Generations*, United Nations University, Tokyo, and Transnational Publishers, New York 1989, 26–27. Also, Brown Weiss, *'Intergenerational Equity: A Legal Framework for Global Environmental Change* in E Brown Weiss (ed), *Environmental Change and International Law: New Challenges and Dimensions*, UN University Press, 1992, 385.
65 The United Nations 1987; Helm 2015, 39.
66 Hallett et al. 2011, 341.
67 Gammage et al. 2021, 102.
68 Hill et al. 2013, 2.
69 Ibid. 35.
70 *Intergovernmental Agreement on the Environment* (1992) s 3.5.2, from Bates et al. 2019 at 8.75.
71 *Minors Oposa v Secretary of State of the Department of Environment and Natural Resources* (1994) 33 ILM 173 see commentary in Bates 2016 at 317.

72 See, for example, *Anderson and Director-General, Department of Environment and Conservation* (2006) 144 ZLGERA 43.
73 *Acland* at paragraph 1306 quoting from *Adani Mining Pty Ltd v Land Services of Coast and Country Inc & Ors* [2015] QLC *48* at paragraph 38.
74 Watson et al. 2010.
75 Helm 2015, 53
76 Ibid. 83.
77 Hallett et al. 2011, 179.
78 Helm 2015.87.
79 Ibid. 35.
80 Ibid. 50.
81 Centre for Air Quality & Health Research and Evaluation 2013, 4.
82 McNeill 2000.
83 Hallett et al. 2011.
84 Helm 2015, 53.
85 Hallett et al. 2011, 23–24.
86 Australian Government, Department of Agriculture, Water and the Environment 2020; Australian Renewable Energy Agency at https://nationalmap.gov.au/renewables/, 2020.
87 Australian Bureau of Statistics, Environmental Assets 2019.
88 Energy assets contributed 4% to total environmental asset value: Natural gas contributed 83% to total energy assets. Crude oil and condensate combined contributed 15% to total energy assets. Mineral assets contributed 6% to total environmental asset value: Iron ore contributed 58% to total mineral assets. Copper, gold and antimony contributed 30% to total mineral assets. Land accounts for 90% of the total value of environmental assets.
89 The UK Office for National Statistics 2019.
90 Helm 2015, 64.
91 Hallett et al. 2011, 366.
92 Ibid. 367.
93 Ibid. 367.
94 Helm 2015.
95 Helm 2015.132.
96 Ibid.
97 Hawken et al. 2010.
98 Ibid. 143–144.
99 Ibid.
100 Keohane and Olmstead 2017, 145.
101 Ibid. 159
102 Goetzmann 2017, 472.
103 Keohane and Olmstead 2017, 34–35.
104 Ibid. 34–35.
105 Ibid. 34–35.
106 Ibid. 57–59.
107 Ibid.
108 Denniss and Campbell 2015.
109 Hawken et al. 2010, 267.

3

SUSTAINABLE GOVERNANCE

Introduction

The following chapter explores a democratic justice system that affords liberty and human rights to support a sustainable nation or local community. Such a system is more likely to attract investment from outside investors and large institutional investors. At a macro level, these principles reduce country risk for investors. At a local level, such a system encourages good local community governance and social benefits for the community.

This chapter analyses the fourth and fifth principles of the Sustainable Development Proposition (SDP):

> AND the democratic rule of law (DRL) is upheld
>
> $\wedge$ DRL
>
> AND human rights and liberties (HR) are upheld
>
> $\wedge$ HR

Property rights, historical context

A historical analysis of the management of the commons from a western perspective shows a complex and well-managed property rights-based system. Property rights are recognised internationally as a human right.[1] Easements, profit a prendre, stinting and covenants have assisted in the management of common property as well as specific legislation. Other accepted processes for managing common property infrastructure include private ownership, Coasean bargaining and government regulation. Common property management can even be found in modern urban environments, such as common property in a high rise building managed under

DOI: 10.4324/9781003324669-3

body corporate legislation. The principle being that under the right conditions, the cooperative management of common property yields more benefits to the collective as a whole.

Some of these historical approaches allow for significant inclusion of the communities in the management of common property. Many indigenous cultures view property as common and collectively owned.[2] Australian Indigenous groups, for example, share the responsibility for managing the environment, for instance, the process of cool burning which prevents bush fires is managed by the group.[3] In Australian Indigenous communities, the harvesting of natural resources, such as fish, is a cooperative customary activity and the harvests are shared with the family group. The harvests are not traditionally valued in monetary terms but in terms of not taking too much of the resource so as not to deplete the natural stock for future generations.[4]

Elinor Ostrom developed the theory of polycentrism which analysed various elements that were most likely to increase cooperation and sustainable management of shared property and resources in communities.[5] Ostrom looked at how community ownership of property is not a new concept but persists in many indigenous and European cultures.[6] She demonstrated in laboratory simulations of social dilemmas how humans can cooperate to produce long- and short-term benefits and that humans do not universally maximise and value short-term self-benefits over cooperation.[7] She also analysed field experiments in which various factors such as the size of the group, leadership, heterogeneity, market access and percentage of conditionally cooperative members (propensity to cooperate was conditional on the cooperation of others) were indicators of the likelihood of successful cooperation to manage common resources.[8]

Ostrom studied how Swiss farmers share mountain pastures as well as the management of local indigenous village-based water irrigation systems in the Philippines. Ostrom observed that the commons were cared for by communities. She demonstrated that communities develop long-term strategies and norms of behaviour that preserve the sustainability of a resource better than if a higher institution had imposed norms upon the community. She showed that a collective or communal ownership could encourage responsibility through non-exclusive common property rights over shared resources.[9]

Indigenous Australians do not own country, country owns them.[10] This gives rise to a view of "property" which entails collective and cooperative resource harvesting or caring for country collaboratively. There was no system of money or time value of money. When going "on country" people pay their respects "to country" and everyone and everything embodied within country. They call out to country announcing who they are, where they are from, their intentions and their recognition of the special nature of country.[11] This concept of ownership is similar to that found in Ostrom's research but quite different from the western concept of managing the commons. As mentioned above, the western concept of property ownership relates to a bundle of rights and sharing those rights is done through legal instruments such as easements, profit a prendre, stinting, covenants and cooperative corporate structures.

Whether the cooperative ideology can be taken from a village community and applied to a city has been explored in the concept of the "sharing economy." The concept started with metropolitan community gardens and grew into IPOs and domains such as shareable.net.[12] Other well-known examples are, Airbnb™, Uber™, Menulog, Airtasker, Ratesetter, Gumtree, eBay™, UpWork, Freelancer, Fiverr and Amazon™.

It has been proposed that cooperative property structure yields the best results in terms of sustainability of energy supply. Polycentrism, it has been argued, responds to energy-related problems effectively as power is shared between the numerous stakeholders, increasing the likelihood of cooperation and resolution:

> These include the involvement of individuals that think in the long-term and see resources as important for their own achievements; the availability of reliable information with minimal transaction costs related to its collection; open and frequent communication among stake-holders regarding costs and benefits; effective rule enforcement and provisions for forcing compliance (such as sanctions); and predictable and gradual changes to rules and enforcement when they occur.[13]

Other scholars have questioned the public policy application of such social economic theory. For instance, Andreas Goldthau asks the following questions with reference to polycentrism as it is applied to energy policy in communities:

> At what point does a project become 'too big' so that potential benefits from economies of scale or cooperation become over-whelmed by cost overruns and challenges? Why do energy megaprojects continue to hold such allure despite their inherent drawbacks? Should the solutions to the energy security challenges identified here match the scales at which they occur, or should solutions transcend them?[14]

Professor Helm from Oxford University propounds that social norms that would sustain a polycentric system are context-specific.[15] This is one of the reasons that case study analysis is used in this book. Experts and central authorities, Helm argues, are needed to make decisions and set regulations.[16] However, this opposes the decentralised cooperative ideology that many communities prefer.

The aim of this guide is to provide a balance between these two models so that maximum consumer and investor uptake can be achieved. A variety of successful models that range from community centred, and not for profit, to investor centred profit based and government funded will be explored. Each different successful model has been applied to a separate community context and the variety is dependent on the enterprise design choices at the community level. These investment models will form part of the case study analysis.

Financial, legal and property development innovations have solved issues of energy and infrastructure supply throughout history. Importantly, where a financial, legal or development solution exists, physical conflict can be avoided.

Units of money play a virtual role rather than a physical role and have done so since ancient Mesopotamia when silver became a unit for expressing the value of different kinds of goods in a single dimension. Money became a tool of thought for making transactions.[17]

> The discovery of calculation (logismos) ended civil conflict and increased concord. For when there is calculation there is no unfair advantage, and there is equality, for it is by calculation that we come to agreement in our transactions.[18]

Financial and legal technology emerged in ancient history as tools to record quantities, verify promises and deal with abstraction in time and space. Inventions such as compound interest and loans with long-term maturity made future development available in the present.

> Technology requires genius to invent it, but it also requires capital investment and legal protection. Infrastructure requires financing, legal protection. Intellectual property protection allows entrepreneurs to keep experimenting while their peers work a steady job. If an entrepreneur faces expropriation by the State or private enterprise of their innovation, they have little incentive to invest in the human capital required. Capital markets and the protection of intellectual property rights can serve as cofactors in sustaining entrepreneurial motivation and capital investment.[19]

Historically, financial and legal innovation has led to infrastructure development, especially energy infrastructure,

> Although European investors were pushing their governments to look after their foreign investments, the stocks and bonds issued by emerging market nations were voluntary. Governments and companies around the world floated loans on the London Exchange of their own volition. A vast number of the securities traded in London paid for modernization, new technology, and infrastructure. The railroads, canals, tramways, and electrical grids in North America, Russia, South America, China, and Africa were built with European investor money.[20]

Goetzmann illustrates in his seminal work, *Money changes everything: how finance made civilization possible*, the history of conceptual tools, such as financial contracts, mortgages, equity and debt instruments, commercial courts, merchant law, private corporations, banks and banking systems, financial planning, models of economic growth, the mathematics of compound interest and empirical records to analyse price trends through history. These are all technological innovations that made infrastructure development achievable in difficult circumstances throughout history.[21] For example, in the Ancient Roman context, the early debt and credit structure of funding initially enabled infrastructure builds in the early cities. This then became impossible to maintain in times of economic downturn. Additionally, the influence of the senators in the lending process made it impossible to build infrastructure independent of power struggles among senators. It was the invention of the corporate structure that put shareholders on an equal footing. Benefits such as per share profits being paid out proportionately and the ability to publicly trade shares to raise capital meant that investments could weather economic downturns and avoid unwanted influence from senators.[22]

Townsend also illustrates how financial innovation can enable infrastructure investment, "Municipal bonds, like a residential mortgage, allow the time horizon for return on investment to be stretched to match the useful working lifetime of the infrastructure."[23]

Social sustainability depends on governance at both the macro and the community level and is varied in its enterprise design. This aspect of the SDP should remain flexible so as not to discriminate against or impose external world views on local rural and indigenous communities. As will be shown in the case studies below local governance structures are chosen by individual communities based on factors and characteristics such as local environment, individual preferences, funding requirements and cultural background.

Australian Indigenous communities have distinctive forms of governance with an emphasis on elders as leaders. Governance practices include decisions by consensus, customary law and "the Dreaming." The most successful contemporary governance institutions are those based on "informed consent, where traditional nodal leaders are respected and empowered, where the local views about cultural legitimacy are taken into account, and where external agencies engage through supportive and facilitating approaches"[24] The governance choices are often community orientated. For instance, 67% of Australian Indigenous entrepreneurs volunteered their time to the local community.[25] And a majority of community-owned Indigenous enterprises are motivated to achieve income for not only themselves but also for their community.[26]

Community driven planning is especially crucial as a success factor in small indigenous communities. There are two noteworthy categories of enterprise design planning, country-based planning and strategic cross sectoral planning. Country-based planning refers to planning that happens on country without interference from external government departments. This only occurs where indigenous people have tenure over the land that doesn't require negotiating with other parties that hold tenure over the same land. The majority of planning occurs in cross sectoral circumstances where various parties some with commercial and some with government interests negotiate in relation to projects on the same parcels of land.[27]

Australian Indigenous people often manage resources through Indigenous Land Use Agreements (ILUAs) which are brokered with multiple other parties such as mining companies, pastoralists, governments and energy companies and are varied across projects and communities. "Indigenous governance is most successful for ILM (indigenous land management) where Indigenous people start it themselves through informed consent, traditional leaders are empowered, the local views are taken into account, and external agencies engage through supportive and facilitating approaches."[28]

Cultural, country-based indigenous-driven planning is identified as crucial for the success of land management projects in the area of "caring for country." For example, instead of allowing outside planners to drive decision-making, "ILM organisations face challenging roles in reconciling cross-cultural encounters. Many are extremely fragile, under-resourced and without access to effective long-term administration, governance and infrastructure support and systems."[29] Indigenous protocols, governance structures and priorities enabled local people to drive decision-making. Indigenous knowledge of ecological practices was incorporated with success.[30] "Country-based planning refers to a process in which Indigenous peoples identify their aspirations and strategies across the whole of their traditional territories, unconstrained by the tenures that are recognised by governments."[31]

Furthermore, indigenous led decision-making has enabled more sustainable projects with less dependency on ongoing government funding and welfare.

> Many successful Indigenous land management activities are part of hybrid economies based on commodities and practices that (i) can be sold in markets; (ii) are underpinned by Indigenous customs; and (iii) are supported to an extent by government investments that stimulate synergies between Indigenous people and market engagement, rather than triggering welfare dependency.[32]

Likewise in rural farming communities, successful projects were found in circumstances where community investment was a key driver of the project. Structures which allowed a large number of entry level community investors were common and successful in those rural areas.

This research is from the investor perspective, looking at how to create communication between communities and outside investors to attract that outside investment to the communities. The aim of the research is to empower communities with knowledge of legal and financial innovation that they can attract investment from non-indigenous private and public institutions.

The method is aimed at empowering indigenous and rural communities to determine for themselves what values they will give weight to when seeking investors. Different communities will have the flexibility to give weight to different investment indicators depending on the direction that they want to take in their enterprise design. The freedom to design and adapt their own projects based on the community decision-making and values is the primary aim of the research.

FIGURE 3.1 A broken traditional wind turbine in the foreground and a new wind generator in the background on an Australian sheep farm. Photo by itpow from iStock.

The SDP developed in this research, working in conjunction with the SCIIs™ has been designed to allow for diversity across communities in rural and regional communities and as such is a flexible matrix. The SCIIs™, in the spirit of OECD sustainability assessments, are tools to assess the viability of an investment from a sustainability perspective. The choices made resulting from that analysis will depend on a variety of context specific factors, such as the availability of resources, institutional capacities and legal structure choices made at governance levels in a variety of communities.[33]

Some communities give weight to financial goals and energy savings, while other communities are focused on social benefits to the local people. By allowing freedom in enterprise design, the case studies show that community-led projects provide successful outcomes. Figure 3.1 shows a rural Australian sheep farm with a traditional wind generator that has been replaced by new wind turbines.

Economic justice

Rawls's theories of economic justice, a concept that relates to governance as well as the economy, suggest that future generations should inherit the necessities for a decent life.[34] He discusses social cooperation as a foundation of choosing and assigning basic rights, duties and division of benefits. He proposes that in assigning these basic rights, duties and benefits, the citizens choose behind a "veil of ignorance" ensuring that no one is disadvantaged by the choice.[35] In order to make an economically just decision, a person looks at the system from the standpoint of the least advantaged. The just economic decision maximises or contributes to the long-term expectations of the least fortunate group, consistent with the just savings principle and opportunities are open to all under conditions of fair equality.[36] For example, in the "equal pay for equal work scenario" which assumes that women are paid less than men for equal work, the person in the thought experiment imagines that they are not yet born (veil of ignorance) and they have an equal chance of being born as a man or as a woman. The person would be imagining that there is a 50% chance of being born a woman (the least advantaged in this scenario) and so would choose a policy of equal pay for equal work was implemented so that they would be treated with fairness in the workplace whether born as a man or as a woman.

Rawls calls these principles "justice as fairness." This means that principles are agreed to in a situation of fairness, where everyone's initial situation is equal and no one can design principles to favour their situation. Therefore, the laws enacted to protect those principles must be laws that all persons would agree to if they were free and equal.

Rawls denies that this can be achieved through utilitarianism because, in such a system, certain individuals must sacrifice their interests and advantages to benefit the maximum good of the whole. He says that no rational being would behave in such a manner unless they had "strong and lasting benevolent impulses."[37]

Amartya Sen discusses this notion of liberty in relation to community development to describe human rights, access to health and education, freedom from poverty, political freedom and rights.[38] Sen contends that "development requires the removal of the major sources of unfreedoms: poverty, tyranny, poor economic opportunities, systemic social deprivation, neglect of public facilities as well as intolerance or overactivity of repressive states."[39] Lack of public facilities includes clean water, peace and order and electricity infrastructure. Sen explains that community infrastructure development must have an evaluative reason and an effectiveness reason, meaning, firstly, that the development will enhance people's lives and secondly that the people will have free agency to participate in the development though democratic processes, such as input into the making of public decisions.[40]

All of these principles are embodied in law in some form in Australia and a community supported by good governance will be sustainable in these aspects of justice and well-being.

One of the objectives of the SDP is to create sustainable governance and social sustainability in communities. That is, to promote economic justice, legal and human rights and good governance under democratic principles for remote communities in essential infrastructure.

At a micro level, more detailed legal governance structures that provide for fairness and sustainable governance within communities are explored and evaluated in the case studies.

In western democracies, we can usually take as given the macro level sustainable governance that comes with democratic rule, the Commonwealth Constitution and the separation of powers doctrine. The historical evolution of governance that led to this structure gives precedent such that, where a system falls outside of these principles, the power structure within Australia provides many avenues for recourse and amendment of the system to bring it back within democratic principles of fairness and justice, for example, through principles of procedural fairness, natural justice, impartiality, human rights, rules of evidence or preservation of liberty.[41]

Improved economic justice in the terms described above would include an end to energy poverty. The social benefits of improved economic justice that will be identified in the case studies are improved health from refrigerated fresh food, increased safety with lighting at night, construction of health centres and schools in remote regions, increased income, local employment and savings in energy costs.

Democracy

The fundamental basis of economic justice is the democratic rule of law that supports human rights such as the right to property. The main principle of democratic law is that the people have the power to change it.[42] Democracy began in Ancient Greece and has evolved through uprisings of oppressed people throughout history. This is usually recorded in a nation's constitution. Many constitutions were hard won through war, such as the *United States Declaration of Independence* in 1776 and *United States Bill of Rights* in 1789. In the case of the US, oppression of the rights of the property owners and citizens led to the Revolutionary War. The outcome of the war was that democratic rights and freedoms were recognised by law. Founding father, Samuel Adams was a leader of the Revolution. Prior to the Revolution he was liable for treason against the King and was threatened with persecution and hanging. Nevertheless, he led the American Revolution in pursuit of property rights and human rights. The more recent Australian Constitution was passed during the peaceful Federation in 1901, allowing Australia to remain not only a part of the Commonwealth but also self-govern.[43]

There are three arms of power within a democracy, the Legislature (Parliament), the Executive (the public service) and the Judicature (for example, courts and judges).[44] Within a democracy, those arms are separate. This means, for example, that the Prime Minister cannot tell a chief of police what to do in his job and vice versa. A chief of police cannot dictate to a judge what to write in his judgement. A judge cannot tell a chief of police how to manage the operations of his police force.

On a micro scale, the administrative arm of a minister's office, including the legal staff are not permitted to directly instruct or advise the elected ministers. Instead, a strict and transparent method of briefing up or down the chain of command is adhered to. The lawyers will brief the Crown Solicitor and any advice to an elected minister from the Crown Solicitor will be subject to the formal briefing process and kept inviolably on the public record for all members of the public to access through freedom of information, right to information or access to public records archiving procedures. This ensures a transparent, independent and accountable democracy that remains a service to the public. This prevents corruption and waste of public money. Where separation of powers is absent, a sovereign nation would be run by an authoritarian regime or gang.

Parliamentary process

Within a democracy, laws must first be tabled in parliament as a Bill and debated among the various political parties in an open public forum: this is parliamentary process.

A draft Bill must also be checked by various lawyers from different departments, including parliamentary committees, public policy departments and the departments to which the legislation impacts. Public consultation must also be undertaken so that the groups or individuals impacted by the draft Bill have an opportunity to have their say on any amendments to the draft Bill before it is tabled. If the Bill is to go ahead after these checks, it will be tabled and debated in a public forum between the different political parties. If the Bill gains enough votes, it will be passed and then assented. A party that is voted into power by the people in a democracy will have an advantage in voting for their own Bills, because they will usually have the majority. This is not guaranteed as members of the same party will not always agree on law changes. Sometimes no party has a majority and the balance of power is held by minority politicians. Even where a party has the majority and does pass a Bill creating a change in law, this is not the end of the line. If the members of the public are not satisfied with the change in the law, they can vote for a different party in the next election on the basis that the new party will amend or repeal the law that the public disagrees with.

Sometimes a political party will use an "emergency power" to create rules that have not gone through this democratic process. This can result in abuses of rights and freedoms and can result in legal cases being brought against the regime and its followers for breaches of rights and freedoms or abuses of power.[45]

The checks and balances involved in democratic parliamentary process will usually identify and rectify any breaches of democratic rights and freedoms before the Bill is passed into law.

Rights and freedoms

There are many rights and freedoms within democratic sovereign nations. Those rights and freedoms are based on hundreds and sometimes thousands of years of common law precedent. Some of the most important examples of rights and freedoms were codified into the *Universal Declaration of Human Rights*[46] after World War II. This Declaration has been ratified by all allied democratic nations. Some examples of the rights and freedoms codified include the right to free speech, the right to life, liberty and security of person (to prevent assault or to prevent unwanted medical experiments or treatments), the right to freedom of movement and residence and the right to leave any country or to return, the right to liberty and the prohibition of arbitrary arrest, detention or exile, the right to an independent and impartial trial (to prevent imprisonment without a jury trial of peers within the separation of powers doctrine), the individual right to privacy, the right to non-interference with the privacy and amenity of the family, including privacy of correspondence and protection of honour and reputation, property rights, the right to freedom, the right to equality, prohibition of slavery, prohibition of torture, prohibition of inhuman or degrading treatment or punishment, the right to recognition as a person before the law, the right to asylum, the right to peaceful assembly, the right to freedom of thought, conscience, religion and expression, the right to employment, the right to education, including the parental rights to choose the kind of education that shall be given to their children, the right to equal treatment before the law, the right to remedy for breach of rights. These rights and freedoms as well as many more not mentioned here could be discussed in great detail; however, this is a topic that requires many more words than are allowed in this book.

Within a democracy, there are different branches of law that are designed to ensure that rights and freedoms are upheld. For example, administrative law is a branch of law that is used to review the actions of judges, politicians and other decision-makers within a democracy. This area of law can provide recourse for abuses of power, breaches of natural justice and procedural fairness. Some examples of where a decision-maker can be in breach include

1. Exercising lawful policy but exercising it in an unlawful way, for example, by failing to take account of the merits of a situation;
2. Exercising lawful policy but doing so without discretion or by acting at the behest or under the dictation of another's discretion;
3. Failure to take into account the rules of natural justice; and
4. Exercising power in such a way that the result of exercising that power is uncertain.

The law of equity was formed by the ecclesiastics in response to the oppressive rule of King Henry VIII who created widespread poverty and unjust subjugation of his people. It is an important area of law within democracy as it overrides legislation in most cases. One example is the fiduciary. A fiduciary duty in equity arises, for example, between an employee and employer or principal/teacher and student or church leader and follower, where one party occupies a position of influence over another and has undertaken to use that position for the benefit or welfare of another.

There are many duties you have as a fiduciary, including the duty not to use undue influence upon the person under your influence, for example, in pressuring or convincing the beneficiary to sign a contract.

The legal practice of tortious liability also prevents abuse of power that can undermine a democracy. For example, courts have held that

> it would be most disastrous to the public generally … for you would simply be governed by the strongest, and whoever was the strongest would

> prevail ....The law is that if a person commits a wrong of that sort he is liable to be mulcted in damages for that wrong, and if he commits a wrong of that sort under circumstances of insult and contumely the person so wronged is entitled to exemplary damages.[47]

In criminal law, a person accused of a crime is presumed innocent and is entitled to an independent court and jury of their peers. They are entitled to procedural fairness, which encompasses the rules of evidence, and to legal representation telling their side of the story to the court, thus ensuring their right to a fair trial. A person cannot be arbitrarily imprisoned in a democratic society: this means a person can only be deprived of liberty if convicted by an independent court and jury under this process.

In a democracy, individuals have the right to own and alienate property on their own behalf or in conjunction with others. This is an important right and freedom and is in opposition to the communist and socialist ideal of communal or government ownership. In most democracies, public utilities can be owned by the government or be privatised, depending on the particular circumstances. However, individuals have a right not only to own their own property but also a right not to be arbitrarily deprived of their property. Property rights are numerous and complicated and would be the subject of a separate text.

It is important to understand the democratic principles of a system on a macro scale so that when community level governance is designed those fundamental principles are upheld within those communities. This ensures that large institutional investors, such as superannuation funds, from western democracies invest in those community projects. For instance, it is more difficult to gain funding from an institutional investor, if human rights are being abused in the community that is seeking investment.

It is also important to remember democratic principles when creating regulatory systems for the emerging renewable energy market. In particular, the competition law regulations to be discussed in the next section impact the attractiveness of investment in a community.

Political economy

Energy infrastructure is dependent in large part on government regulation which is nation specific. In particular, competition law in relation to monopolies of infrastructure is relevant. When competition law, company regulation and energy market regulation is passed, it sometimes has unintended and unforeseen market consequences such that leaders should be prepared to monitor the market effects of legislation and make amendments where necessary. For instance, overregulation to enforce competition or conversely overregulation that encourages the creation of monopoly markets domestically can incentivise companies to invest in alternative less regulated offshore markets. However, too little regulation creates uncertainty in the law and therefore uncertainty for investment in onshore capital assets. This could also lead to investment capital being taken offshore. Therefore, a balanced approach should be aimed for.

In a democracy the energy market can be regulated in three primary ways:

1. Centralised, State government owned. The market is controlled by monopolies and duopolies and State-owned vertically integrated assets. The Federal government can provide oversight.
2. Decentralised, democratised, private investment, with a focus on consumer protection and competition law.
3. Balance and partnership between these two systems.

In the first scenario, the investor is less likely to risk capital investment due to the high probability of State government control. The investment money is tax revenue driven. However, when investment dries up, so do taxes, ending in a downward spiral of economically unsustainable infrastructure. Affordability for the consumer is necessary for engagement in the market design, especially as public tax payer funded assets are often sold for a profit to private companies under this market design. This can in turn cause country risk and an increase in energy network access prices for local people.

In the second scenario, overregulation can be a barrier to private investment, especially where private investors are prohibited from efficiently managing all aspects of the business due to prohibitions on vertically integrated companies. In this scenario, investors are reliant on the strength of their contracts and business advisors or community trust and cooperation.

In summary, when it comes to regulation, a balance must be struck between the two extremes, they are:

1. Too little regulation; and
2. Too much regulation.

Both extremes cause consumer dissatisfaction and consumers opting out of the market as well as a disincentive for investment and a disincentive for cooperation from public, donation or private investors.

A further risk with energy market regulation is uncertainty of political will. This occurs with market instruments such as feed-in-tariffs (FiTs). When FiTs are constantly changing, investors do not have certainty in relation to the return on investment time frames for infrastructure, especially in small scale projects where power purchase agreements are not available for negotiation.

A fine balance must be struck and monitoring of the market post regulatory reform is recommended as well as frequent stakeholder consultation and amendments to regulations where necessary.

Alternatively, community infrastructure projects, funded and structured in a variety of economically diverse and innovative models (private, donation, public and hybrid), can provide non-network solutions in a sensible, balanced way. This allows infrastructure to be built where needed in "energy islands" that can be more affordably connected to national grid systems at a future time where need be. As stated by Dr Finkel, this reduces upfront network costs, increases participation in the market and flattens demand peaks. Non-network solutions also create resilience through diversity as they can operate as stand-alone or connected systems when designed to do so.[48] The case studies analysed below demonstrate that this model is often sustainable; economically, environmentally and technologically. Economically and legally diverse structures can provide innovative solutions to energy infrastructure supply for communities.

Much of the regulatory work towards renewable energy uptake was done in Australia between 2018 and 2020 with updates being passed in Federal legislation to revive the Retailer Reliability Obligation and provide a National Energy Guarantee as well as improvements to the Australian Competition Law. Additionally, significant funding has been announced at Federal and State levels for renewable energy investment. In the decade previous to these improvements, multiple reports were published investigating the National Electricity Market (NEM) regulations and renewable energy investment conditions in Australia. These changes will likely attract even more renewable energy investment in Australian markets.[49]

Accordingly, the Australian energy marketplace is reviewed in the next section in relation to political economy and legal governance.

Energy law and policy

A fair regulatory framework that is stable and not subject to constant revision gives investors the confidence to put their money into projects.[50] An unstable policy and regulatory framework in the renewable energy industry has adversely impacted Australia in the past. In 2014, Australia's ranking for renewable energy investment attractiveness fell from 11th to 39th place. This was due to uncertainty regarding carbon pricing legislation (which was repealed as it incentivises large corporations to shift their pollution to more vulnerable economies) and Renewable Energy Target regulations.[51] Since then, Australia has gone from 39th place to ranking 4th in 2020 for renewable energy investment attractiveness.[52] Internationally, the US has surpassed China for the first time, ranking number 1 in attractiveness for investment in renewables. The primary reason is reported as being: financial innovation that looks beyond subsidies.[53] In the US, 9.1 GW of wind was added, bringing the total to 105.6 GW and 13.3 GW of solar was built, bringing the total to 77.7 GW. Incentives such as the Production Tax Credit, Investment Tax Credit, safe harbour rules and FiTs are driving the investment.[54] It is not only governments that are driving investment in renewables. Many funds are switching investments to renewable, sustainable and ethical investments. Crédit Agricole is just one example among many international funds that has announced it is withdrawing from coal power and shifting its trillion dollar asset management business to clean energy sources.[55] A 2019 PwC survey on the investment priorities of 750 institutional investors and 10,000 retail investors ranked environmental sustainability concerns as third, below risk-return but above fees, relationships and operational capacity.[56] Over $30.7 trillion worth of assets was invested sustainably in developing countries in 2018. In Australia, Japan and New Zealand, growth in sustainability increased approximately 308% a year between 2014 and 2018, amounting to about $12 trillion.[57]

Market power

Following the initial research into renewable energy various grants have been awarded to research the policy and regulatory aspects of the energy market. In relation to market power, the Australian Energy Regulator (AER) in the *State of the Energy Market Report* (SEMR Report) (2009) found that companies, who own peaking units, could exploit high demand events such as peak demand caused by extreme heat.[58] Despite this vulnerability in the system, in relation to misuse of market power in the electricity market there have been no adverse rulings against market participants for their behaviour in the electricity market. The ACCC follows the precedent set in the decision *AEMC, Final Rule Determination: Potential Generator Market Power in the NEM*, 26 April 2013. In that decision, it was held that the price spikes that occurred under certain weather conditions were *not* the result of abuse of market power in Queensland and Victoria.

The Federal government has recently taken the risk of abuse of market power seriously proposing strong interventionist powers for the Commonwealth.[59] In 2018, the ACCC finalised a report on market power in the electricity industry, "Restoring electricity affordability and Australia's competitive advantage: Retail Electricity Pricing Inquiry." In particular, the ACCC, "for the time being, has rejected the idea that a market power mitigation rule should be introduced, suggesting that wholesale power price increases are less a factor of misuse of market power than a lack of competitive constraints."[60]

Further to this, the Federal government drafted extensive regulations[61] to resolve the competition issues reported by the ACCC in 2018.[62] In addition to competition issues, blackouts, price rises and affordability issues were reviewed by the ACCC.[63] This has resulted in widespread legislative reforms to consumer law which could benefit consumers, especially vulnerable electricity customers.[64]

Some of the changes to the electricity market include:

- A post 2025 Market design for the NEM.[65]
- The Consumer Data Right has been applied to the electricity sector.[66]
- Increased competition in the marketplace.[67]
- Retailer reliability obligations.[68]
- Embedded Network Service Providers regulations.[69]
- Standardised billing.[70]

- Removal of excessive penalties applied to vulnerable customers, for example, without paying their bill before the due date.[71]
- Simplifying market offers so they can be compared.[72]
- Increased transparency for new generation development projects.[73]
- A National Energy Guarantee.[74]

The legislative reforms are extensive and apply predominantly to large-scale projects. However, some of the proposed amendments directly impact vulnerable consumers positively. Much feedback has been provided by key stakeholders and law firms. Standard practice is that the Federal government will consider this feedback, make amendments where necessary and then table the final draft.

It is encouraging to see that the above list of changes to the regulations has largely implemented Dr Finkel's recommendations to drive investment in the energy sector. Dr Finkel's recommendations are summarised below.

- A long-term emissions reduction trajectory;
- Stable energy policy;
- A Clean Energy Target;
- Increased resilience and security;
- Reliability of supply;
- Rewards for consumers;
- Non-network solutions; and
- All existing large electricity generators will be required to provide a binding three years' notice of closure. This will signal investment opportunities for new generation and give time for communities to adjust, including provisions made for transitional employment to the renewable energy sector.

Given the scrutiny that Australia is under with regards to carbon emissions policies, it is prudent to note that any aim to decarbonise the electricity industry must be done responsibly as predicted by Dr Finkel below:

> Given the regional location of many coal-fired generators, the transition of these employees must be well planned. A notice of closure requirement for generators, as discussed in Chapter 3, would facilitate a well-planned transition. New jobs could be created by building large-scale VRE generators in those regions.[75]
>
> A 50 year lifetime limit on coal-fired generators, for example, would allow existing coal-fired generation assets to operate to the end of their expected investment life. A number of large coal-fired generation owners … have previously announced their intention to not invest in new conventional coal assets, while Origin and AGL have also announced specific timeframes by which they plan to divest from coal-fired generation assets. Together, these three companies are the majority owners of more than 60 per cent of the

> privately-held coal-fired generation capacity in the NEM. This suggests that a 50 year lifetime limit would not be at odds with the sector's intentions or expectations. At the same time, a lifetime limit would improve planning information and certainty for the sector.[76]

The main recommendation above that pertains to this research is non-network solutions. These are necessary to counteract the power affordability crisis, to stabilise the grid, increase resilience in the power supply, incentivise communities to connect to the grid where economically viable and provide services to remote communities.

The installation of smart metering and off-peak vehicle charging can forecast peak demand and flatten electricity demand peaks. Smart metering can reduce the vulnerability of the system to exploitation by peaking units.[77] However, there are also several risks with smart metering due to the vulnerability of consumers in contracting with retailers, especially consumers in marginalised remote communities. Competition and diversity in the marketplace for smart metering should reduce the likelihood of consumers being exploited by smart metering companies. The recently passed Commonwealth Consumer Data Right legislation assists in protecting consumers.[78]

The priority recommendation for any new technology such as smart metering that can adversely impact vulnerable consumers is that the basic principles of contract law are upheld. It is prudent for businesses to comply with these principles to ensure they are not exposed to legal action for unfair contracts. Some examples (not exhaustive) of those principles are:

- Smart meters are voluntary;
- Consumers are reasonably able to opt in or out of contracts; and
- Mutual intention to be bound by all of the terms of the contract is clear.

Power affordability: investment in grid infrastructure

"Under reforms introduced in 2015, the AER can remove inefficient investment from a network's asset base if the network overspent its allowance, to ensure customers do not pay for it."[79] Previously overinvestment in grid infrastructure has led to unaffordability in the electricity market. This weakness in the market was uncovered by the AER in the *State of the Energy Market Report* (SEMR Report) (2009). Additionally, the ACCC report drew attention to unethical practices causing overinvestment in poles and wires.[80] Overall 43% of costs that are passed through to consumers were estimated by the ACCC to be network costs.

The ACCC report found that the current regulatory framework incentivises businesses to overinvest in grid infrastructure. This is because under the regulations a business can recover network charges based on a level of return (the weighted average cost of capital) on the networks Regulatory Asset Bases (RAB). Therefore, the more money the business spends on the network, the higher the network charges that the business is allowed to pass on to the consumer.[81] The ACCC Report states that this has incentivised an overinvestment in grid infrastructure. It is this increase in network charges that is in turn resulting in electricity unaffordability and incentives for customers to disconnect from the grid.

The AER has a limited ability to review the efficiency of capital investment in the grid. Overall 50% of the $4 billion worth of investment in network assets in the NEM made between 2005 and 2017 was considered excessive by the ACCC. These costs were passed on to the vulnerable consumer as network charges.[82]

This overinvestment in poles and wires has occurred in many countries. International Economist, Droeg, observes a global trend, "Investment that continues to be expended on antiquated systems represents wasteful investment in failing infrastructure and directly competes with much needed resources for new technology. This also diminishes the reserves needed to build a new and survivable electrical base."[83]

The former Chief Scientist, Dr Finkel, recommended that the overcharging of network capital costs could be addressed by the Australian Energy Market Commission by assessing alternative models for network incentives and revenue-setting, including a total expenditure approach. Dr Finkel also recommended non-network solutions as being the key.[84]

In 2019, the AER was involved in a dispute in relation to $5 billion of network charges to customers. Following the review and appeal process, the network companies (some of which are foreign government corporations) are now directed to repay the excess revenue back to the consumers.[85]

Some areas of Australia, such as Queensland, require the pricing for cities and regional areas to be equitable. The means that the network charges for city dwellers are subsidising the transmission lines out to the agricultural areas. This is seen as a fair system as the agricultural belts support the city food supply. Additionally, as a response to the frequent power outages in the 1990s in Queensland, Energy Queensland is now required to maintain a certain percentage of expenditure on infrastructure to prevent power outages from occurring under strict reliability standards.[86] Energy Queensland must maintain a fine balance between maintaining the required level of expenditure under State regulations and not overspending under the Federal energy regulations. The AER approved less distribution investment for Energex and Ergon in Queensland in 2020 than in the previous regulatory period, despite requests from those companies to increase capital expenditure. This framework of checks and balances is aimed at reducing the retail costs to consumers.[87] If the AER sets the rate of return too low, a network business may not be able to attract sufficient funds to invest in assets needed for a reliable power supply. If the rate is set too high, the network businesses have a greater incentive to over-invest.[88] Overall a balance must be met each year. Looking to the future, the current regulatory framework of checks and balances will likely result in equitable pricing of network charges.

Power affordability: non-network solutions

Dr Finkel proposed the following solutions to overinvestment in the grid:

- Network businesses should be rewarded for providing the services that consumers want, such as non-network solutions;
- The regulatory framework should be adjusted so as to stop favouring capital investments, resulting in too much expenditure on network assets and too little expenditure on non-network alternatives;
- The regulatory framework should encourage network businesses to utilise new technologies where they are cheaper than building poles and wires;
- Non-network solutions could include purchasing services provided by DER or utilising individual power systems or micro grids as alternatives to a traditional grid connection.[89]

Non-network solutions include infrastructure such as micro grids, hybrid systems and stand-alone systems. Dr Finkel stated that energy governance should be focused on attracting, "Increased investment in large, medium and small-scale variable renewable electricity (VRE) generation capacity and micro grids."[90] He explained,

> Micro grids, stand-alone power systems and DER can strengthen the NEM's resilience ... the COAG Energy Council should direct the Australian Energy Market Commission to undertake a review of the regulation of individual power systems and micro grids so that these systems can be used where it is efficient to do so while retaining appropriate consumer protections. The Australian Energy Market Commission should draft a proposed rule change to support this recommendation.[91]

The AEMO, the Australian Energy Council, the Energy Users Association of Australia, General Electric and the Clean Energy Finance Corporation all made submissions to Dr Finkel stating that "delivering a secure power system with a high VRE penetration is technically and economically feasible." Dr Finkel points out,

> While a number of studies have found that there are no technical barriers to a high VRE penetration in the Australian context, it is important that there are efficient and equitable approaches for providing security services, and capturing their benefits. Work is already underway by AEMO and the AEMC in this regard. Solutions must be implemented on a region-by-region basis, so that costs and needs are balanced.[92]

He explains,

> Demand management is critical in managing potential disruptions of electricity due to extreme weather conditions. Micro grids, stand-alone power

> systems and DER can strengthen the NEM's resilience as well as function as a grid resource for faster system response and recovery. Inter-regional connections can help regions draw power from unaffected areas and make the power system more resilient to extreme weather and other natural hazards.[93]

The following relevant observations were made by Dr Finkel in this regard:

> Modelling conducted by Energeia for the *Electricity Network Transformation Roadmap* found that new regulatory arrangements would be required to allow innovative service delivery options for up to 27,000 new rural connections between now and 2050. Almost $700 million could be saved by supplying these connections, usually farms, with an individual power system, but current regulations would require the use of a more expensive conventional grid connected service.
>
> The Energy Council's Energy Market Transformation Project Team published a consultation paper in August 2016 on how these systems could be regulated and whether there is value in regulating them under the National Electricity Rules and National Energy Retail Rules.
>
> The AEMC is also currently considering a rule change request from Western Power that seeks to remove some of the barriers to distribution network businesses using individual power systems as an alternative to a grid connection, particularly in rural areas.[94]

Super grids are costly and logistically difficult to plan and build, requiring significant investment.[95] Small communities can build micro grids fairly easily and cost-effectively. This is why entities such as Western Power (in Western Australia) and Powerlink (Queensland) are funding non-network solutions (such as micro-grid solar and battery storage) to supplement electricity supply as the population grows.[96]

This does not mean that micro grids will be off-grid indefinitely. The long-term plan would involve Australia's inland transmission grid proposal, which plans to connect 900 km of transmission line through rural Australia.[97] This could be implemented in a similar way to the Islands strategy in Europe. The Islands strategy is an investment strategy employed in Europe to overcome the expense of grid connecting separate cities and communities. The first stage is to create linkable networks, also known as islands. The aim is to connect these energy islands over time. At stage one, self-sufficient energy islands are built. At stage 2, the interconnecting infrastructure is built. This means less upfront investment on national grid connection is needed. Therefore, less network costs are passed onto the consumer alleviating the unaffordability issues.[98] The reliance on long-distance transmission networks constitute up to half of total hardware construction costs in conventional systems.[99]

Transparency in the planning stage of the inland transmission grid would be essential to the success of the plan. This is because if stand-alone systems intend to be grid connected in the future, they must have the specifications to meet the future transmission lines or have the capabilities to be upgraded to meet future specifications. Engineers need this information when designing the systems. This is a normal procedure, similar to that used when State governments issue a regional plan under planning instruments.

Virtual grids are another solution that is currently being trialled by some private companies as well as the AER and are discussed in a subsequent chapter.

Power affordability: sharing economy solutions

There are many commercial solutions that employ non-network solutions. These have been analysed in the case studies below. Reviews to current Commonwealth legislation that support this are being considered. The main recommendations being:

- Households must have strong opt in/opt out rights in relation to smart metres and contracts with retailers to ensure competition in the market; and
- Proposed market reforms which would allow customers to directly interact with suppliers in the NEM in a "two-sided market" much like Uber or Airbnb.[100]

The primary benefits of sharing economy solutions are:

1. Flexibility in purchasing and feed-in times, flattening the peaking in the marketplace and reducing costs for the consumer;
2. Reduced pressure on the grid;
3. Decreased overall costs; and
4. Incentives for households and small businesses to invest in smart meters and, smart appliances and distributed energy resources, thereby, increasing efficiency and participation in the NEM.[101]
5. Privacy protection for consumers.

Summary

In summary, the SDP relies on the democratic rule of law that in turn ensures that human rights and liberties are upheld. A system will only be sustainable long term where freedom and justice are ensured. We need not dwell here on the history lessons in relation to authoritarian and fascist regimes that have sought to impose principles on citizens in a mandatory way by use of brute force and restriction of liberty. Those systems are invariably fallible and cannot be sustained long term. However, it is also prudent to remember not only the large political regimes that have failed in this manner but also the multitude of "little" human rights that make up a democracy, such as the right to privacy that is supported by the Australian Privacy Principles and various international agreements between Australia and other allied nations, the right to own and alienate private property or the right to quiet and peaceful enjoyment of one's home and family. The democratic justice system is such that if one of these multitude of rights is trodden on in pursuit of a technological advancement or environmental action, it is likely that the citizens living within that system may cease to support it and as a result the system will become unsustainable long term.

Importantly, for this guide, when human rights and democratic principles are upheld, communities are more likely to achieve social benefits and are also more able to attract investment from large institutional investors. Therefore, the following two propositions are included in the SDP.

AND the democratic rule of law (DRL) is upheld

$\wedge$ DRL

AND human rights and liberties (HR) are upheld

$\wedge$ HR

Notes

1 United Nations, Universal Declaration of Human Rights, Article 17.
2 Gammage et al. 2021.
3 Ibid.
4 Hill et al. 2013.
5 Ostrom 1993.
6 Andersson and Ostrom 2010.
7 Ostrom 1993.
8 Vollan and Ostrom 2010.
9 Ibid.
10 Kennedy et al. 2018, 17–18.
11 Ibid.
12 Hollis 2013, 151.
13 Andersson and Ostrom 2010 from Sovacool 2014. *Energy Research & Social Science*, 16.
14 Goldthau A. Rethinking the governance of energy infrastructure: scale, decentralization and polycentrism. Energy Res Soc Sci 2014; 1:134–140, from Sovacool. 2014. *Energy Research & Social Science*, 17.
15 Helm 2015, 119.
16 Ibid. 120.
17 Goetzmann 2017, 15.
18 Ibid. 94.
19 Ibid. 197.
20 Ibid.
21 Ibid. 135.

22 Ibid. 135.
23 Townsend 2013, 289.
24 Hill et al. 2013, 35.
25 Ibid. 2.
26 Ibid. 6.
27 Hill et al. 2013, 51.
28 Ibid.10.
29 Ibid. 3.
30 Hill et al. 2013, 51. Citing the following case studies: Miriuwung-Gajerrong protected area co-management in the east Kimberley (Hill 2011; Hill et al. 2008b), numerous small Indigenous communities in central Australia (Walsh & Mitchell 2002), joint management endeavours at Mootwingee National Park in New South Wales (Lane & Hibbard 2005) and cultural and natural resource planning in the wet tropics (Larsen & Pannell 2006; Worth 2005), and commentary in Moorcroft et al. (2012). See also, Wunambal Gaambera Traditional Owners (Hill et al. 2011b; Wunambal Gaambera Aboriginal Corporation 2010).
31 Hill et al. 2013, 51, citing Smyth 2008.
32 Hill et al. 2013, 2.
33 Ibid.
34 Rawls 1971.
35 Ibid.
36 Ibid.
37 Ibid.
38 Sen 2001.
39 Ibid. 3.
40 Ibid. 4.
41 Judge Chisholm and Nettheim 2002.
42 Ibid.
43 *Commonwealth of Australia Constitution Act 1900.*
44 See for example, the Commonwealth of Australia Constitution Act 1901 or The Constitution of the United States of America, As Amended, July 25, 2007.
45 See for example, *Yardley v Minister for Workplace Relations and Safety* [2022] NZHC 291 [25 February 2022].
46 United Nations, at www.un.org/en/about-us/universal-declaration-of-human-rights
47 *Lumley v Gye* EWHC QB J73 (1853) 118 ER 749 (1853) 2 Ellis and Blackburn 216.
48 Finkel 2017, 72.
49 See for example, COAG Energy Council, Energy Security Board (2019) *Post 2025 Market Design for the National Electricity Market (NEM)* at www.coagenergycouncil.gov.au/publications/post-2025-market-design-national-electricity-market-nem, accessed 7 February 2020; *The Treasury Laws Amendment (Consumer Data Right) Act 2019*; *National Electricity Rule proposed amendments, 2018*; Final Determination – Interim Reliability Instrument Guidelines Retailer Reliability Obligation July 2019; National Electricity (South Australia) (Retailer Reliability Obligation) Amendment Bill 2018; *National Gas (Capacity Trading and Auctions) Amendment Rule 2018.*
50 Australian Energy Regulator (AER) 2009.
51 Axup 2017.
52 Ernst & Young Global Limited 2019.
53 Ibid.
54 Ibid.

55 PWC 2020.
56 Ernst & Young Global Limited 2019.
57 Ibid.
58 Australian Energy Regulator (AER) 2009.
59 Collyer et al. 2018.
60 Downes 2018. *ACCC wants changes to the National Electricity Market.*
61 *Treasury Laws Amendment (Prohibiting Energy Market Misconduct) Bill 2018.*
62 ACCC 2018.
63 ACCC 2018.
64 *Treasury Laws Amendment (Prohibiting Energy Market Misconduct) Bill 2018* which amended *the Competition and Consumer Act 2010.*
65 COAG Energy Council 2019
66 *Treasury Laws Amendment (Prohibiting Energy Market Misconduct) Bill 2018* which amended the *Consumer Data Right Act 2019.*
67 Ibid.
68 Final Determination – Interim Reliability Instrument Guidelines Retailer Reliability Obligation July 2019; National Electricity (South Australia) (Retailer Reliability Obligation) Amendment Bill 2018.
69 *Treasury Laws Amendment (Prohibiting Energy Market Misconduct) Bill 2018* which amended *the Competition and Consumer Act 2010.*
70 Ibid.
71 Ibid.
72 Ibid.
73 *National Gas (Capacity Trading and Auctions) Amendment Rule 2018; National Electricity Rule, Proposed Amendments 2018.*
74 *Treasury Laws Amendment (Prohibiting Energy Market Misconduct) Bill 2018* which amended *the Competition and Consumer Act 2010.*
75 Finkel 2017, 73.
76 Ibid. 97.
77 Ibid.
78 *The Treasury Laws Bill Amendment of the Consumer Data Right Act 2019.*
79 Australian Energy Regulator (AER) 2009, 136.
80 ACCC 2018.
81 Downes 2018. *ACCC wants changes to the National Electricity Market.*
82 Ibid.
83 Droege 2009, 6.
84 Finkel 2017, 25.
85 Australian Energy Regulator 2020, 127.
86 Ibid. 133.
87 Ibid. 141.
88 Ibid. 146.
89 AGL submission to the Review, 5–6, from Finkel 2017, 150–151.
90 Finkel 2017.
91 Ibid.
92 Ibid. 49–50.
93 Ibid.72.
94 Ibid. 154, 338. From Energy Networks Australia, www.energynetworks.com.au/energeia-modelling-roles-and-incentives-microgrids-and-stand-alone-power-systems, accessed 3 June 2017; COAG Energy Council, *Stand-alone energy systems in the Electricity Market: Consultation on regulatory implications*, 2016; AEMC, www.aemc.gov.au/Rule-Changes/Alternatives-to-grid-supplied-network-services, accessed 3 June 2017.

95 Powerlink 2018.
96 Powerlink 2018, 33–45; Dickers 2016.
97 TransGrid 2020.
98 Droege 2009, 30.
99 Ibid.
100 Downes 2018.
101 Australian Energy Market Commission, *How Digitalisation is Changing the NEM: The Potential to Move to a Two-sided Market*, Information Paper, 14 November 2019, at www.aemc.gov.au/sites/default/files/20191/How%20digitalisation%20is%20changing%20the%20NEM.pdf, accessed 7 February 2020.

4

SUSTAINABLE ECONOMY

Introduction

A common-sense approach dictates that an increase in standard of living as population increases is much preferable to a decrease in standard of living as population increases. It is a myth that increasing population causes a decrease in standards of living due to pressures on natural resources and resource scarcity. In actual fact, as cities and urban centres in western society increase in population, the standard of living in those cities correspondingly improves. This is what drives the trend towards urbanisation, that is, the migration of populations from remote rural areas to populated city centres and the social benefit of an increased standard of living that ensues.[1]

Economic sustainability for a project is about ensuring a system is self-sufficient economically for an indefinite period of time or until the project is decommissioned without the need for continued outside grants, donations or funding to keep the infrastructure project viable. Various models that support economic sustainability are demonstrated and evaluated in the case studies.

Economic sustainability is also about maintaining sustainable yields of natural resources. This is explored further in the Sustainable environment chapter, including, analysis of how to value natural capital resources so that sustainable yields can be monitored.

Governance for communities at a macro scale, such as democratic governance explained above, must be balanced with local community governance and social benefit goals at a community level. Local community governance means including the individual community members in the enterprise design decisions of infrastructure projects and allowing self-determination in development projects. Community-centred governance is explored in the case studies. This also contributes to economic sustainability in some community contexts.

DOI: 10.4324/9781003324669-4

FIGURE 4.1 A rural farm in the foreground powered by a white modern wind turbine in the background. Photo by Pixabay from Pexels.

Social benefits arise from an increased standard of living which is a consequence of the reduction in energy poverty for remote communities.

One important social and economic benefit of renewable energy projects is the employment opportunity being created for communities.

Below, Aboriginal elder, Bess Nungarrayi Price,[2] demonstrates the needs in her community, the full quote is provided so as to ensure the entire context is captured,

> My mother and father were born in the desert. They taught me the old, sacred Law that our people lived by. It worked when we were living in tiny family groups taking everything that we needed from the desert. It was strong for sacred business and for marriage. Men had the power of life and death over their wives. Young girls were forced into marriage with older men. There was no law for property except that everything must be shared. There was no law for money, houses, cars, grog, petrol or drugs: we didn't have any. The only way to punish was by beating or killing the law breakers. We had no army, police, or courts and only our family to defend us. Law Men used magic to heal or harm and kill. Now we have a new law that is not sacred and doesn't believe in magic. We have property, houses, cars, grog, drugs, pornography. We live off welfare or we need to get a job to live. We still share everything. We can't say 'no' to kin even if they waste our money or destroy themselves with it. Too many men still want the power of life and death over their wives. We don't plan for the future, budget or invest. Too many of our kids are at terrible risk. There's too much drunkenness, feuding, sickness, suicide, unemployment and despair. Our kids are not being educated properly in either law. The old Law was not about human rights. It was about unconditional loyalty and obedience. Wise old people tried to make sure that there was justice.

> But even they couldn't deal with grog, drugs and violence. It all happened too quickly. We still respect and honour our ancestors and want to keep our culture. But my people are confused. If they go the blackfella way they break whitefella law, if they go whitefella way they break blackfella law. Our young men are caught in the middle, that's why they fill up the jails. My parents understood that things were falling apart. My mother outlived eight of her eleven children. They understood the difference between the letter of the Law and its spirit. We now need to change the letter of our Law to keep its spirit alive. We need to do this ourselves but with the support of governments and our fellow citizens. The time for shouting slogans and waving placards is over. We need to do some tough thinking and honest debating.[3]

The above quote demonstrates a call from Australian Aboriginal elders to increase the standard of living in their communities and thereby create social benefits such as, gaining a consistent rule of law, increasing the safety of women and children and improving employment opportunity. Employment is a significant aspect of increasing the standard of living in developing communities and increasing economic participation. As Bess Nungarrayi Price, "I've got a job. And I think without employment, no jobs for our people, and education, Aboriginal people don't exist, more or less. Jobs create stability, family togetherness. But that's not the way Aboriginal people live."[4]

Indigenous labour force participation rate has only increased marginally in the last few decades. It is about three quarters of that for non-aboriginal people. Community Development Employment Project (CDEP) participants are the major part of the labour force in remote areas. In some communities when the CDEP figures are removed from census employment figures, there is near total unemployment in those communities. To compare, non-indigenous unemployment declines in remote areas.[5]

It is important that the local employment opportunities as they form part of the Sustainable Community Investment Indicators are considered in infrastructure projects. Other benefits that will be explored in the case studies include population increase, with Indigenous people moving from outlying isolated areas into communities with solar power, access to refrigeration for fresh food and medication, improved health and hygiene through the use of electric stoves and washing machines, improved access to education through school of the air and improved safety with street lights.

Social benefit is relevant to this chapter on Economic Sustainability. This is because if the standard of living increases over time as the population increases, this leads to a myriad of social benefits that are also economic benefits, such as energy security, income and employment. The social aspects of a community energy project are those aspects that have increased the standards of living for the community, and, thus, they are intrinsically dependent on the economic sustainability of the project.

Social benefits of an increased standard of living

What follows is an analysis of the sixth, seventh and eighth principles of the Sustainable Development Proposition:

> AND economic growth (EG) increases in line with the population (POP)
>
> Λ (POP $\leftrightarrow$ EG)
>
> AND technology (T) improves
>
> Λ T
>
> AND the standard of living (SL) increases in line with population (POP) growth
>
> Λ (POP $\leftrightarrow$ SL)

Sustainable economy: when technology improves

When technology improves, the standard of living can also improve, even where population in a community is increasing. However, nowadays, some people live in an attitude of fear due to the issues discussed above pertaining to natural capital balance sheets. They are afraid of population growth and afraid of scarcity of resources.[6] As will be explained below, technology can and does find solutions to supporting increasing populations while ensuring that the standard of living for those populations continues to improve.

Dr Common, ecological economist, explains that sustainable yields from renewable resources must be improved with technological progress if sustainability is to endure long term. Technological progress with a Cobb-Douglas function demonstrates this principle of sustainable yields from renewable resources.[7]

Yields are usually placed in the following categories:

- Direct benefits or utility to the observer.
- Aesthetic, but not utilitarian in use.
- Aesthetic beauty and a significant eco service, for example, the provision of a natural form of flood defence or defence against erosion.
- A core energy source, resource, or food source.[8]

Dr Helm explains, "These categories directly or indirectly produce value to people, and can be broken down into ecosystems, species, freshwater, land, minerals, the air and oceans, as well as natural processes and functions."[9]

Professor Common has conducted modelling of the relationship between population, affluence of an economy and impact on the environment and technology (The IPAT Identity). This modelling demonstrates that given sufficient technological progress, the Cobb-Douglas function demonstrates that ongoing economic growth is possible with a growing human population using a sustainable yield from a renewable resource. Sufficiently fast technological progress can keep the growth per capita income increasing over an extended period of time.[10] The key to the modelling is that where the technologies or means by which goods and services are produced can be improved at the same rate that population and affluence increase, the yields from the renewable resource will be sustained. Technology includes financial and legal tools as well as renewable energy technology.

Thus, a combination of technological progress and sustainable yields from renewable resources can lead to sustained economic growth for communities even where population increases and consumption increases.

The key to the success of the Cobb-Douglas function is that increased cycles of consumption are avoided by improvements in technology, for example, digital technology replacing paper-based technology could improve the efficiency of the renewable forestry resource even where a population is larger and consuming more information.[11] Another example is the southern region of Australia that supported a much larger Indigenous population due to the increased fish stocks, through the technology of trapping fish in pools to breed the additional renewable food resources and through stone huts that did not require deforestation or timber resources to build.[12] This technology caused an increase in standards of living, an increase in renewable natural capital as well as an increase in population.

However, as Townsend explains, any gains in efficiency can lead to a rebound in consumption or an increased cycle of consumption. A reduction in the cost of the resource can spur consumers to use that resource. Urban planners call this the Jevons paradox. In transportation, for example, building the new road does not reduce traffic for long but due to the new capacity, "the opportunity cost of driving falls, spurring drivers who would never have ventured onto the previously clogged road to sally forth."[13]

The Economic Optimism model begins with the premise that the only general conclusion that one can reach from examining long-run economic growth models is that there is no general conclusion.[14] Although resource scarcity can constrain global economic growth, technological change has demonstrated time and again its ability to outpace the influence of resource constraints.

James Lovelock, the scientist who discovered the damage CFCs were doing to the ozone layer and helped organise the world-wide ban of CFCs and consequent repair of the ozone layer, explained that the Earth is a living, breathing entity made up of living systems and individual life forms[15]. If this is so, as the only known sentient beings, we are necessarily the responsible custodians of the living Earth and must use our unique technological, intellectual and social ability to fulfil that role.

Population policies

When the standard of living improves in a country, the birth rate of that country begins to decline and eventually becomes "negative birth rate" meaning the population is not replacing itself but is decreasing. For instance, more couples will begin to have one or no children in developed countries with increased standards of living. This is good news for pressure on resources, however, it is a concern for economic growth as population increase drives economic growth. Consequently, governments in developed countries with negative birth rates have instituted policies to address this issue.

Australia is one of many developed nations that is working to address the long-term and continued decline in fertility rates that have occurred in the last 50 years among developed nations. The current fertility rate in Australia is 1.66 per woman which means that population growth is below replacement and has been since 1974.[16] In many European countries, there is a negative fertility rate due to the ageing population and decrease in births. In Australia, there is a ministerial task force aimed at addressing work – family issues through the implementation of various employment-related strategies and policies. This is because the impact that a declining population growth has on the labour force and the economy is negative.[17] Legislative reforms such as mandatory reporting of work and family balance policies for organisations with greater than 100 employees, the expansion of legal protections for those with family responsibilities, requirements for female participation on boards and availability to request part-time work up to a child's second birthday are some of the incentives to encourage fertility as well as labour force participation and improvement in education opportunities for Australian women.[18]

Therefore, too much population growth can be remedied by increasing the standards of living for women thereby decreasing the birth rate. Standards of living can be increased by improving equal opportunities for women within a society so that women have access to birth control and have freedom within their relationships to choose how many children they have. Also, there must be an increased standard of living that thereby increases the survival rate of their children. When the standard of living increases to the point of creating a negative birth rate, this can be managed by work-life balance and other employment and policy incentives or immigration. Thus, as the standard of living increases and the fertility rate decreases, incentives for women to participate in economic opportunities must be increased through policy reforms.

This book will demonstrate how the use of renewable resources in a Sustainable System, even with population growth, can lead to sustainability of resources for a region, and sustainability in all aspects: financial, technological, environmental, democratic governance and social.

By comparison, strict sustainability such as population control can place limits on human rights and freedoms.[19] Developing countries in particular are vulnerable to strong sustainability that impacts human rights and freedoms. If, for example, the growth of China or India's economy were halted to zero growth in order to protect resources, this kind of logic could keep China's less fortunate in the same state as the 1970s after the Great Famine and the Cultural Revolution, with atrocities such as the one child policy (an estimated 70 million deaths resulted from famines at this time in history).[20]

Strict sustainability that impacts human rights and freedoms is not needed to maintain a Sustainable System. It is only when participants fail to rationally understand the consequences of their natural resource extraction that systems collapse.[21] Therefore, thoughtful decision-making that increases the standard of living in line with population growth, enabled by the Sustainable Development Proposition, can lead to Sustainable System builds.

The increased standards of living that come with economic growth can continue even where population increases. Additionally, as standards of living increase, population begins to follow a negative trend as negative fertility rates accompany increased standards of living in communities; therefore, the fear of population increase can be averted by focusing on increasing standards of living.

In particular, technological advancements can increase standards of living as well as help to ensure that natural capital supports populations even where populations do increase in communities.

Thus, the sixth, seventh and eighth proposition of the Sustainable Development Proposition (economic growth increases in line with population AND technology improves AND the standard of living increases in line with population growth) are added to the previous propositions outlined earlier. In combination, the resultant propositions describe a Sustainable System (SS).

Summary

Infrastructure and technology that is environmentally sustainable, as outlined above, is the first step to a sustainable community. If sustainable technology is to be built in communities, funding is needed and funding must be sustainable. For example, if a loan structure is used, the loans must be serviceable. Technology requires genius to invent it, but it also requires capital and operational investment and legal protection. Therefore, sustainable governance and financial innovation is also required.[22]

Where economic sustainability is achieved through improving standards of living in line with population growth, the social benefits, such as energy security, improved health services and employment and business opportunities, are also achieved. Thus, economic and social sustainability are intrinsically linked through providing increased standards of living for populations. Therefore, the following propositions are included in the Sustainable Development Proposition.

AND economic growth (EG) increases in line with the population (POP)

$\wedge$ (POP $\leftrightarrow$ EG)

AND technology (T) improves

$\wedge$ T

AND the standard of living (SL) increases in line with population (POP) growth

$\wedge$ (POP $\leftrightarrow$ SL)

As community infrastructure becomes more complex, financial and legal innovation are needed to keep up with the fast moving societal and technological changes. Development is dependent on financial tools to move value through time and to restructure the myriad economic risks when building communities.[23] This financial and legal technology are explored in the case studies and outlined in the Attracting Investment chapter.

The case studies provide a comprehensive analysis of the primary property economic solutions to infrastructure supply in remote regions, demonstrating the most successful ways to attract investment for renewable energy projects. The case studies are selected from a range of renewable energy technology solutions. Those technology solutions are reviewed in the next chapter.

Notes

1 Hollis 2013, 151.
2 Member of the Commonwealth Government's Advisory Group on Violence against Women.
3 Johns 2011.
4 Paul 2019.
5 Johns 2011, 20–21.
6 Hallett et al. 2011, 98.
7 Common and Stagl 2005, 224.
8 Helm 2015, 3.
9 Ibid.
10 Common and Stagl 2005, 225.
11 This is called the Jevons paradox, after William Stanley Jevons, from Hallett et al. 2011, 320.
12 First Footprints 2013.
13 Townsend 2013, 317.
14 Keohane and Olmstead 2017, 234.
15 See Lovelock 1987, 2006.
16 Australian Bureau of Statistics, "Births, Australia".
17 McDonald 2003.
18 Ibid.
19 Dikotter 2010, from Helm 2015, 49.
20 Ibid.
21 Ibid. 322.
22 For example, Intellectual property protection allows entrepreneurs to keep experimenting while their peers work a steady job. If an entrepreneur faces expropriation by the State or private enterprise of their innovation, they have little incentive to invest in the human capital required.
23 Goetzmann 2017, 520.

5

SUSTAINABLE TECHNOLOGY

- Wind;
- Hydro;
- Biogas;
- Solar PV including silicone technology, affordability, invertors and anti-islanding and passive solar homes;
- Energy storage;
- Virtual grids;
- Electric vehicles (EVs);
- Waste plastic oil;
- Micro grids; and
- Virtual grids.

Introduction

Non-network renewable energy solutions (off-grid or stand-alone micro-grid) are best suited to the remote regions in terms of affordability and efficiency of infrastructure builds. As stated by former Chief Scientist, Dr Finkel, this reduces upfront network costs, increases participation in the market and flattens demand peaks. Non-network solutions also create resilience through diversity as they can operate as stand-alone or connected systems when designed to do so.[1] The case studies analysed below demonstrate that this model is often sustainable; economically, environmentally and technologically.

Non-network systems are the perfect starting point for upgrading energy systems. They can encourage innovation, experimentation and diversity, out of which the most technologically, economically and environmentally Sustainable Systems (SSs) will emerge.

Non-network renewable energy solutions are made up of a mix of 100% renewable technologies, defined as wind, solar with storage and demand management

DOI: 10.4324/9781003324669-5

technology, small-scale hydro and biogas.[2] The case studies have been chosen within this range of renewable energy technology.

Dr Finkel recommends non-network solutions as being necessary to counteract the power affordability crisis, to stabilise the grid, increase resilience in the power supply, incentivise communities to connect to the grid where economically viable and provide services to remote communities.

Non-network solutions include infrastructure such as micro grids, hybrid systems and stand-alone systems. Dr Finkel stated that energy governance should be focused on attracting "Increased investment in large, medium and small-scale variable renewable electricity (VRE) generation capacity and micro grids."[3] He explained, "Micro grids, stand-alone power systems and Distributed energy resources (DER) can strengthen the NEM's resilience …."[4]

Network solutions such as large-scale solar can also be beneficial for rural communities. For instance, two large-scale solar farms in rural Australia are reviewed in the case studies and show good profitability and benefits for the rural area.

The renewable energy solutions explored in the case studies are wind, solar, hydro and biogas. See Figure 5.1 for an example of a rural wind energy solution. The sustainability of those technologies as well as other technologies that are not as sustainable is outlined below.

FIGURE 5.1 White wind turbines in an agricultural field. Photo by Damian Barczak from Pexels.

Small-scale wind

Stanford University in the US has modelled global wind power capacity and reports that seven times the global electricity demand could be satisfied through wind energy alone using only 20% of global potential locations and 6.9 m/s wind speed at 80 m above the ground with standard 1.5 MW turbines. European countries have been reconfiguring their national grids based on this study since it was published in 2005.[5] In particular Germany and Denmark procure a large percentage of their electricity from wind energy.[6] Municipal-scale renewable energy has also been built in California.[7] Offshore wind energy is highly efficient and has potential to provide high-value return on investment.[8] Small-scale wind projects in remote and regional areas are likewise attracting investment in agricultural regions, such as rural Australia and Europe. This is explored in the case studies.

Small-scale hydro

Micro hydro is small-scale hydro that doesn't have a dam. It has a lower environmental cost due to the small intake.[9]

The viability equation:[10]

$$\begin{aligned} &Head(\textit{fall in meters}) \times Water(L/second) \times 9.8(gravity) \\ &\quad \times Efficiency(\textit{assumed } 70\% \textit{ approximately}) \\ &\quad = Power(watts) \end{aligned}$$

Hydro viability is dependent on the appropriate watercourse being available for development within a community and so is location specific. Micro hydro is explored in more detail in the Upper Yarra Community Power Case Study. In that case study, the site chosen had been previously used for hydropower before electrification a century earlier.

Biogas

Some regions, such as national parks with thick tree cover, are not suitable for solar. In these scenarios, biogas can be used as an alternative renewable energy solution.

India, Indonesia and China have been developing rural biogas digesters for agricultural communities since the 1960s, aiming, firstly, to reduce energy poverty for rural communities, and, secondly, to reduce the pollution caused by the burning of biomass. More recently, Germany has started manufacturing modern biogas plants for large-scale energy production.[11] Across Europe, the sector has grown, more than tripling to 17,400 biogas plants in 2015. Germany is the leading biogas country in Europe producing 23 TW of biogas energy.[12]

Biomass burning is the traditional way of cooking. In many rural communities, wood or wood chips are burnt for heating or cooking. The smoke from the wood burning causes lung disease for the women and children who predominantly do the cooking in those communities. Additionally, when the surrounding areas are deforested to provide firewood, soil erosion occurs, destroying the agricultural base of the communities. Cooking smoke is the highest health risk factor above smoking and high blood pressure in India and the south-east Asian region. Air pollution in that region from conventional wood stoves is responsible for 4 million deaths each year. The cost to the national healthcare system (not reflected in the price of wood or energy) is between $212 billion and $1.1 trillion for the region. Women and children are most vulnerable to these hazards as well as the hazards of spending up to 4 hours a day collecting fuel. In addition to the pollution, there are many reported cases of women and children sustaining injuries while collecting fuel.[13]

The issues noted above are the reason biogas plants have come to replace traditional biomass burning. Biogas is collected in a small-scale plant so that bacteria can anaerobically digest the biomass. The methane is then burnt for cooking. There are two primary types of biogas plants:

1. Covered effluent ponds for liquid waste where the biogas is piped for processing; and
2. Engineered digesters in fermentation tanks. The process is controlled with heating and cooling and adding bacteria.[14]

The feedstock for the plants ranges from livestock or human effluent to organic landfill and wastewater or decaying vegetation or food waste.[15]

Biogas captured in digesters typically contains 55–20% methane, 20–20% carbon dioxide and trace gases such as toxic hydrogen sulphide and nitrous oxide. Methane has 21 times the power of carbon dioxide as a greenhouse gas. Capturing and burning it can theoretically be better than letting it be released into the atmosphere; however, more environmental studies would be needed if this was to be conducted on a large scale.[16] This is because emissions from the burnt biogas also need to be considered.

As the burning of gas still creates pollution and warming, this technology is best used where solar and wind are not viable. For example, in national parks, where tree coverage prohibits solar capture and forest residue is collected for disposal, a biogas plant could be used instead of solar panels. Also, industries where large amounts of methane are emitted, such as the agricultural industry could benefit from additional revenue for the disposal and processing of the methane waste product.

The small-scale 2–3 m^3 plants are used for home cooking and heating. Larger industrial plants, such as those manufactured in Germany can be used by businesses or communities.[17]

Solar PV

In Australia, the solar hot water industry grows by about 30% per annum.[18] Solar PV has been used to produce electricity since the 1950s.[19] Over time, the affordability increases.

Solar energy grew by leaps and bounds in the 2010s. According to Wood Mackenzie, global annual solar installations grew more than sixfold this past decade, from 16 GW in 2010 to 105 GW in 2019.[20]

In the meantime, multi-silicon solar module prices dropped from over $2 per watt to just over $0.20 per watt in 2019, making a 90% price reduction and driving the investment in solar.[21] It is the fastest cost reduction out of any electricity generation technology over this period.[22] Comparatively, the overnight capital cost required to build onshore wind and natural-gas-combined cycle power plants in the US decreased by 38 and 2%, respectively, between 2010 and 2018,[23] whereas the cost to build combustion turbine plants has gone up by 11%. The same unfavourable economics have led to very few new coal plant constructions.[24]

The International Renewable Energy Agency (IRENA) reported the global levelised cost of onshore wind generation fell by 35% between 2010 and 2018.

> Over the same period, it reported the global levelised cost of large scale solar PV fell by 77 per cent. In Australia, the Commonwealth Scientific and Industrial Research Organisation (CSIRO) and Australian Energy Market Operator (AEMO) in December 2019 estimated a levelised cost of electricity (LCOE) in 2020 for large scale solar PV and onshore wind of around $50 per megawatt hour (MWh). They forecast the cost of onshore wind will continue to reduce marginally to 2050, but the cost of large scale solar PV will reduce by almost half in that time. The substantial cost reductions observed in wind and solar technology have made these renewables the lowest cost option for new build generation. The CSIRO found the cost of those technologies is significantly lower than construction costs for new black coal and brown coal generators (and significantly lower than the cost of coal generation with carbon capture and storage). The lifecycle costs of wind and solar generators are now becoming competitive with the operational costs of the current fleet of conventional generators.[25]

Globally the solar energy market has increased from US$13.6 billion in 2006 to US$69.3 billion in 2016. Solar energy is booming globally with semiconductor manufacturers (Applied Materials, GE and Sharp) as well as start-ups entering the field.[26]

The installed solar capacity Australia-wide is increasing rapidly. In late 2000, Australia had approximately 400 kW of installed grid-connected solar power. By 2013, the same amount was being installed on a single building. Installed capacity doubles on average about every 10 months. The installed household solar power in Australia is around 10 GW with large-scale farms contributing around an additional 1,824 MW.[27] Germany and China have the largest capacity of installed solar power. Australia has the potential to reach 54 GW of grid-generated capacity if suitable Australian homes and other commercial roofing are utilised to capacity. This is without even considering large-scale facilities.[28]

Dr Finkel states,

> An increasing proportion of investment in new generation assets comes from individual consumers. In the NEM, consumers have installed more than 1.44 million rooftop solar photovoltaic systems. AEMO forecasts that by 2036 the annual electricity generation from rooftop photovoltaic solar will increase by 350 per cent from current levels.[29]

It has been estimated that the wind and solar power market could be 20 times as large as it is today and that this alone would encourage further investment and further market growth.[30] In Figure 5.2 a solar farm has been built on an agricultural holding in Australia thus providing an additional source of income for the farmers.

Depending on the material used, 1 kW per square metre of sunlight energy can be converted by solar cells into 3–5 W of power in full sun. Different solar arrays can be used to produce different results. Depending on the type of cell and the colour band of the sunlight converted, the efficiency can vary significantly. The angle of the cells, and whether those cells are smart cells, for example, whether they can rotate to collect the sun as the sun moves, affects the efficiency. Solar PV, unlike solar thermal, works better at lower ambient temperatures. Typically, there will be a 0.5% change in efficiency per degree Celsius increase.[31]

These efficiencies continue to increase at a rapid rate. Solar PV modules can be made using monocrystalline PV cells, polycrystalline PV cells or thin-film technology (photovoltaic material between two sheets of glass or plastic). Monocrystalline is the most efficient, has the best temperature coefficient and is the most expensive. Polycrystalline is less efficient but cheaper. Thin-film technology

FIGURE 5.2 A solar farm in agricultural Australia. Photo by Kelly L from Pexels.

has a low-temperature coefficient, making it useful for hot climates. It is also versatile.[32] Australian professor, Martin Green and his team at the University of New South Wales (UNSW) have held the world record in silicon solar efficiency for the last 38 years, beating the likes of Elon Musk in the annual World Energy Prize.[33] Professor Green invented the PERC solar cell (Passivated Emitter and Rear Contact) which is used in 85% of the world's solar cell production. Sales of PERC solar cells are expected to grow to more than US$3 trillion by 2040.[34]

New research into silicon cell technology is advancing rapidly. Overall, 50% of all silicon solar panels incorporate technology developed in Australia.[35]

Germany has been investing in solar significantly since 2000. The number of people employed in the PV industry has surpassed those working in fossil-fuel-fired power. Employment has been created in manufacturing, engineering and installation. The solar market is worth over €8 billion annually to Germany.[36]

Recent research conducted by the Royal Meteorological Society, UK, and the CSIRO in relation to "Potential impact of solar arrays on regional climate and array efficiency"[37] included experiments in relation to the potential benefit/impact on climate, microclimate, environment and the agricultural belts surrounding large-scale solar sites that could be built across Australia to take advantage of the 58 million PJ of energy that Australia receives from the sun annually.[38] The study found that the climate over and around solar arrays can be modified dependent on the design of the arrays. In the surrounding environment, rainfall can be decreased by approximately 30–70% or alternatively increased, air temperatures could be decreased or increased by factors of up to 10°C, solar irradiance could be increased by between 5 and 20% and solar power yields can be increased by up to 25%.[39]

The study shows how the array sizes, locations, orientations and surface albedo can be manipulated to control the regional climate surrounding the large-scale solar farms. Depending on how the solar farms are designed, this can have a significant benefit to the environment and should be taken into account when designing large-scale solar in regional areas. It is important that surrounding farms and biodiverse environments are considered when designing the large-scale solar farms so that rainfall and temperature are not impacted adversely but are instead impacted to the benefit of the surrounding environment.

Potential follow-up studies include simulating solar cell arrays in further realistic situations, for example, panels elevated above the surface, precise calculation of exchanges of heat from ground to cells and cells to atmosphere.[40] In the Sahara, large-scale solar systems have been studied to increase rainfall and subsequently increase the growth of vegetation.[41]

Solar cell technology that reabsorbs luminescent signals instead of reflecting them is also being developed. Much as a leaf absorbs light without reflection, so too can solar cells. NASA is currently investigating technology to maximise solar cell reabsorption.[42] This could reduce the impact that large-scale solar has on the temperature and rainfall of the surrounding environment. Although in desert regions, such as Australia and the Sahara, large-scale reflective solar arrays can be used beneficially to increase rainfall and revegetate those regions if designed to do so.

Affordability

The affordability of solar technology depends somewhat on the government rebates and grants available. It also changes with the feed-in-tariffs (FiT) that are available when power generated is sold back to the grid. Sometimes the FiT can work as a market incentive for non-network suppliers to connect to the grid and contribute to network costs and maintenance. If managed poorly, the FiT can cause consumers to disconnect from the grid, either because network costs are artificially inflated or because the FiT amounts are unreliable, upsetting consumers. In 2013, the Council of Australian Governments (COAG) agreed to a set of national principles which state,

> Market participants should provide payment for exported electricity which reflects the value of that energy in the relevant electricity market and the relevant electricity network it feeds into, taking into account the time of day during which the energy is exported.[43]

When this principle is honoured, the FiT should work across all Australian jurisdictions. In some countries, FiTs can be locked in for the first year or more of a project so that the income is reliable.

Many micro grids and off-grid systems cannot sell power to the grid as they are non-network infrastructure solutions. However, in some regional communities, infrastructure will be connected to the grid to provide additional income for farming and regional communities. Even off-grid micro grids have been shown to have a short payback period of around two to three years as demonstrated by the case studies. Most solar panels will last approximately 25 years, according to the warranties.[44] So even without a FiT, solar can provide free energy for most of its life cycle.

From a broader perspective, a renewable energy economy would save on average $9 billion per year on power sector fuel costs and $11 billion a year on transport fuel costs. It is estimated that a renewable investment boom could bring in $800 billion of investment to Australia.[45]

A renewable energy certificate (REC) trading scheme (RET) can also provide an additional income. The scheme was intended to help create a fair marketplace. However, there is rarely complete certainty around the risk level of an investment when predictions of future markets are involved in the risk/reward analysis. State and Territory governments also offer grants and subsidies to help renewable investment.

According to UK economist Droeg, a combination of FiTs and non-network solutions providing a return and supported by a dramatic reduction in fossil fuel subsidies will stabilise the renewable energy market.[46]

The FiT system is an effective and equitable way of encouraging renewable electricity generation.[47] It is also the most effective way to encourage people to stay on grid and support the grid infrastructure. The German Federal Environment

Ministry reports that a FiT is low cost, generates employment, improves research and development and innovation, increases manufacturing and mitigates fossil fuel price rises.[48]

Inverters and anti-islanding

Where a solar system is required to operate as a stand-alone system and the owner also wants to take advantage of a FiT by selling electricity to the grid or using the grid as a backup system, consideration in relation to inverters and islanding is prudent.

Using the correct inverters ensures the islanding safety feature that allows inverters to automatically shut down in case of a blackout. This is so that electrical lines workers are not at risk, since, in the event of an outage, they need to be able to reliably and safely isolate poles and wires from electricity to perform grid maintenance.[49]

It should be noted that some unscrupulous energy providers have been taking advantage of customers by overcharging in relation to this service. Some distributors have been falsely claiming that all of their customers must pay fees for so-called anti-islanding testing. In the recent Western Australian Case, *ACN 158 148 951 Pty Ltd v Prout* [2019] WASCA 59 (April 9, 2019), an energy distributer was prosecuted by the Commissioner for Consumer Protection for falsely charging customers $350 for "mandatory anti-islanding testing." The distributer misled consumers into believing that anti-islanding testing was mandatory under the law.

The Victoria State Government states, "The Certificate of Electrical Safety (CES) is a requirement of the *Electricity Safety Act 1998* and *Electricity Safety (Installations) Regulations 2009* for all prescribed electrical installation work." And that, "The distributor will organise for any metering upgrades/remote configuration and will generally conduct an anti-islanding test before allowing the panels to be connected to the grid."[50] There is therefore no need for any mandatory testing or related fees. In Queensland and New South Wales, the use of a generator or battery is usually recommended to supply the home or business during interruptions to supply, such as blackouts.[51] In summary, industry practice is to install an inverter that can handle blackouts, backup generators and backup battery storage systems. For example, an electrician would do the following:

- If off-grid, install solar panels, backup batteries and an inverter that can handle a generator as well as the backup battery system.
- If on-grid, install solar panels, an inverter that shuts down power to the grid during a blackout and an inverter that allows batteries and/or generator to kick in during blackouts.

When islanding, the electrician will know when and how to install different kinds of inverters for safety. A grid tie inverter is needed to match the DC power from the solar PV module to the AC power on the grid. A good inverter should have several maximum power point trackers connected to each string of solar PV models, to dynamically adjust the varying current and voltage. Micro inverters are now available allowing mains wiring direct to the panel in a parallel connection, and they are tolerant of shade and flexible for orientation because each micro inverter can be optimised independently to various surfaces of the roof; however, they can cost more than a string invertor.[52]

Passive solar homes

When designing renewable energy projects for indigenous communities, it is prudent to investigate the quality of housing in the community. If the housing is due for upgrade and investment then the design of the energy system should be built into the new housing design.

Passive solar homes are just one of many self-sufficient home designs that can be investigated by indigenous communities or the various State departments of housing and communities. Net zero energy buildings have been built in many communities. Designing zero net energy housing for remote communities is an area of investment that has not been fully realised in the construction industry.

Energy storage

Current battery technology and the recycling of lithium-ion batteries are regulated under occupational health and safety regulations and environmental regulations in most jurisdictions.[53] Battery recycling has been identified as an opportunity for new industry investment.[54] One option is that customers could receive cash-back when they recycle the car or home batteries at an approved manufacturer or recycling plant. This is a highly competitive market in Norway.[55] Some car manufacturers are opting to retain ownership of the car batteries so that consumers will be required to bring the car battery in for recycling at the end of the contract. From a fair contract perspective, this practice should result in a car sales price that is reduced by the value of the battery that the consumer doesn't own and also accounts for the consumer's time and effort expended in returning the battery to the manufacturer.

New storage methods are being investigated as an alternative to lithium and lead acid batteries. Many different types of battery storage technologies are currently being developed, for example:

1. Graphene supercapacitors with a high electrode surface.[56]
   This is a non-toxic approach to electrochemical storage. The technology uses a graphene nanosheet under a low temperature. It is a low-temperature vacuum-assisted thermal process developed for graphene-reduced oxide.

> The outstanding electrochemical performance of the VAG supercapacitors meets the power, energy, and durability requirements of many applications

> that current energy storage technologies can't. The graphene synthesis process developed in this work can be easily scaled up for high volume manufacturing. The appealing properties of this graphene supercapacitor technology could herald a new wave of electrochemical energy storage.[57]

These vacuum-assisted supercapacitors have an energy density capacity that can be improved. Scientists are experimenting with the conducting polymers and gases used in the vacuum to improve the energy density capacity.[58]

An Australian company has recently patented one such graphene-based battery storage technology. The battery comprises an aluminium-graphene-oxygen chemistry and is said to be safer, more stable, lighter and have a superior energy density that lithium-ion batteries.[59]

2. NASA battery experimentation
   New battery technologies are being explored by NASA to replace the lithium-ion batteries. Regenerative fuel cells that store energy within a thermal-vacuum environment are being tested because the current battery technology is insufficient for lunar exploration missions in shaded regions.
   NASA is currently testing a range of battery solutions as follows:

   a. X-57 SCEPTOR
   b. Convergent Aeronautics Solutions –M-SHELLS
   c. Convergent Aeronautics Solutions –L-ION
   d. High-Temperature Tolerant Batteries –LLISSE
   e. Modular
      i. Li-Air batteries have high theoretical energy densities and potential to leverage on-board oxygen systems, but electrolytes are limiting.
      ii. Investigating novel "electrolyte engineering" concepts to enable Li-Air batteries with high practical energy densities, rechargeability and safety.
      iii. Utilising predictive computation, material science, fundamental chemistry and electric flight testing.
   f. 100 W 10-cell battery, 24 V nominal
      i. Inorganic electrolytes
      ii. Cathode coatings: transparent conducting oxide coatings.[60]

3. Salt
   Figure 5.3 shows a molten salt storage facility in remote China. It has been used as a large storage solution for large-scale solar facilities for some time.[61] It offers a zero-emissions technology solution. It also has a firm capacity and dispatch characteristics. The other benefit is that it is readily available and thus cost-effective. The cooling towers do not require water, which is important for dry areas that usually have high solar and low rainfall or limited water resources, such as outback Australia.[62] This technology has more recently been converted into a small-scale storage solution for rooftop solar.[63]

FIGURE 5.3 A panoramic view of molten salt storage at a solar thermal power station in China (Gansu province, Dunhuang city). Photo by Chuyu from iStock Photo, Getty Images.

Similarly, sodium is also being used in electric car battery innovation in China as it is cheaper and more readily available than lithium.[64]

In summary, it can be seen battery technology and battery recycling industries are innovative, fast moving and prime for investment but must consider the environmental impacts of their technologies. According to Dr Finkel, "Battery storage is poised to be the next major consumer-driven deployment of energy technology."[65]

Virtual grids

Virtual power plants (VPPs) are individually, distributed energy resources that can charge and discharge according to a programme in order to flatten peaks in a grid. Such coordination of the flow of power can assist with grid stabilisation.

VPPs have successfully avoided a blackout in Queensland during a power surge at a coal plant in one instance but in another instance in South Australia underperformed, an issue that was quickly rectified by the Tesla operator who remotely upgraded the capabilities of the DERs operating in the VPP.[66]

Invertors used in the VPPs must meet the new requirements of the AER in South Australia. In particular, certain firmware upgrades that meet ride through standards to ensure that DERs can ride through or disconnect during power system disturbances or "surges."[67]

Virtual grids comprise virtual power plants that can take pressure off the grid by managing individual household solar use and feed-ins.[68] Virtual grids use demand management technology. There are multiple competitive players in this marketplace both domestically and internationally. These range from start-ups[69] to large government-owned enterprises such as Yurika Pty Ltd, a subsidiary of Energy

Queensland. Yurika has developed a virtual power plant that provides the behind-the-meter electricity to embedded networks for the large property portfolios owned by the Queensland Investment Corporation (QIC).[70]

The importance of competition in the marketplace as protected under Australia Law is paramount to ensure that virtual grids provide the most reliable and affordable energy to consumers.[71] It is important that they are legally compliant, in particular with privacy laws. At this point, there have been a number of breaches. However, the AER reports,

> a number of distributors have worked effectively to remediate breaches, and strengthen systems and processes to support compliance. But compliance could still be improved in a number of areas, particularly in separating staff between the distributor and its affiliates, protecting confidential electricity information about the network, and ensuring any shared costs are appropriately allocated between the distributor and an affiliate. However, when breaches have occurred, distributors have mostly communicated promptly with the AER, acted quickly to contain any potential harms from those breaches, and put in place plans to prevent breaches from recurring.[72]

Electric vehicles (EVs) and mobile grids

It has been shown that a shift to electric vehicles (EVs) operating on 100% renewable electricity is realistic. It has been estimated that if battery technology progresses and maintenance costs for EVs are lower than ICEs (internal combustion engines), a shift to electric cars could be achievable.[73] It is estimated that EV sales will overtake internal combustion engine sales by the year 2030.[74] In some countries, it will be much sooner. Between 2013 and 2018, the market share of EVs in Norway rose from 3 to 60% and it is still growing.[75]

This industry's success will depend on:

- EV battery cost;
- Battery power density;
- Availability of charging infrastructure;
- Manufacturing capabilities;
- Customer preference; and
- Legal requirements and incentives to drive investment, for example, in Europe, EVs are tax and road tolls exempt.[76]

Conversely, in Australia, some States are proposing EV road taxes which will disincentivise EV purchases. Taxes on luxury vehicles in Australia also limit consumer uptake. Countries such as the the US, France, Germany, the UK and China now have a subsidy available for EV purchase.[77]

Shifting to 100% electric buses for urban transport was found to be the most sustainable solution for public transport.[78]

FIGURE 5.4 An electric vehicle being charged by a solar- powered charging station in a car park. Photo by Kindel Media from Pexels.

A community that is run self-sufficiently on a renewable energy micro grid can conceivably run all of its transport vehicles with the electricity produced by that micro grid. EVs, when used effectively, can also transform the micro grid into a mobile grid. This is where the EVs connected to the micro grid can feed stored energy back into the grid during peak demand periods or transport stored energy to other separate micro grids nearby if needed. Only certain brands of EVs are capable of this feature.[79] Figure 5.4 shows an EV being charged by solar in a car park.

Innovations such as two-way EV chargers are assisting with this transition to a mobile grid and can be used for rooftop solar storage in households as well.[80]

In the US, the Department of Energy provides low-interest loans for corporate innovation and alternative energy and energy efficiency companies. Tesla was the recipient of such a loan. Since its inception in 2009, the low-interest loan programme has been profitable for the US government.[81]

The AER reported,

> In coming years, customers will increasingly store surplus energy from solar PV systems in batteries, and draw on it when needed. In this way, they will reduce their demand for electricity from the grid. The owners of DER can thus better control their electricity use and power bills, while taking initiative

> on environmental concerns. If DER is properly integrated with the power system, they could also help manage demand peaks and security issues in the grid ....[82]

Tesla and Uber and other companies are also manufacturing electric aeroplanes, ferries, transport vehicles and road trains, many of which are commercially operating. All of these vehicles can contribute to a mobile grid.

The installation of smart metering and off-peak vehicle charging can forecast peak demand and flatten electricity demand peaks. Smart metering can reduce the vulnerability of the system to exploitation by peaking units.[83] However, there are also several risks with smart metering due to the vulnerability of consumers in contracting with retailers, especially consumers in marginalised remote communities. Competition and diversity in the marketplace for smart metering should reduce the likelihood of consumers being exploited by smart metering companies. The recently passed Commonwealth Consumer Data Right legislation assists in protecting consumers.[84] Additionally, the Australian Privacy Principles will ensure that consumer data is handled appropriately.[85]

The priority recommendation for any new technology such as smart metering that can adversely impact vulnerable consumers is that the basic principles of contract law are upheld. It is prudent for businesses to comply with these principles to ensure they are not exposed to legal action for unfair contracts. Some examples (not exhaustive) of those principles are:

- Smart meters are voluntary;
- Consumers are reasonably able to opt in or out of contracts; and
- Mutual intention to be bound by all of the terms of the contract is clear.

Waste plastic oil

Recycling has been gaining momentum as a profitable industry supported by governments worldwide as part of a circular economy solution to pollution and waste products. In particular, plastics have gained attention in the Australian, US and European markets following supportive policy changes. The key objective of the circular economy is to keep valuable resources and materials such as plastics in the economy and prevent unnecessary losses. Targets are set by reference to quality and quantity of output.[86]

The industry has gained more support and attention globally following a series of policy announcements from Asian countries banning the imports of waste-derived materials. Prior to 2017, much of the waste plastics and metals globally were exported to Asian recycling markets.[87] The reports suggest that 210 tonnes of contaminated waste exports are going to be returned to Australia. This trend is known as the global recycling crisis.[88]

Ecosystems are by nature circular. They have a circular metabolism in which every output discharged by an organism then becomes an input that sustains the community as a whole. It is this concept that imbues the circular economy.[89] Recycling is a simple example of a circular economy solution.

Various companies are experimenting with recycling plastic into products, such as furniture, pipes, pallets, posts, bottles, computers, televisions, white goods, commercial equipment, hospital supplies, car parts and printer cartridges. The technology can also be used to make biofuels. Waste plastic oil (WPO) is one such diesel replacement.[90]

The disposal and recycling of waste plastic in the modern circular economy is critical to the long-term sustainability of our natural ecosystems.[91] It is also critical that economies adapt to the circular model in order to compete with modern industrial markets in Europe where circular industrial economies are advancing rapidly, reducing industrial costs and obtaining market share as a result.[92]

Recycled plastics are a key element of this economic growth area. This is demonstrated by the European Union's recent Circular Plastics Alliance declaration.[93]

WPO is an emerging product that has valuable applications in any industry where it can replace diesel as a fuel. A large proportion of remote communities rely on diesel generators for electricity supply.[94] Diesel is used in power generation due to higher fuel conversion efficiency, reliability, durability and torque capability than petroleum engines.[95] WPO is an alternative to fossil-fuel-derived diesel and can be applied in any diesel engine.[96] This makes it a viable solution to the supply of energy to remote communities because diesel is used in hybrid energy systems in many remote regions.[97] Hybrid systems are a combination of solar power and backup diesel generators. Those backup diesel generators could use WPO instead of diesel, making them more environmentally sustainable.

> Bioplastics are also catching on and will soon move to replace the more widely used petrochemical based products in the near future. Adding to the allure of biofuels is the fact they burn cleaner in engines and generate less toxic emissions. This has been widely known and acknowledged within the automotive industry since its inception. Rudolf Diesel, the father of the diesel engine in 1912, actually envisaged his engine running on vegetable oils. Similarly, Henry Ford's Model T was designed to run on either ethanol or gasoline since Ford favoured growing fuel locally. The availability of cheap oil and the 1930s decision to use lead as an additive to reduce engine knocking rather than ethanol effectively put biofuels on the back burner. Today's growing environmental concerns have changed all that .... Toyota is gearing up to manufacture 20 million tons of bioplastics by 2020 hopefully capturing around two-thirds of the global market and generating $38 billion in revenues. The development of new biopolymers and other materials is currently the subject of intensive work.[98]

In 2006, $20.5 billion was invested in biofuels globally, ten years later in 2016, and $80.9 billion was invested, demonstrating the growth potential of this industry.[99]

Farmed biodiesel is no longer supported as an environmental solution as it has created unsustainable farming practices across Europe. Recently the European Union has strengthened criteria for ensuring bioenergy sustainability to curtail farmed biodiesel.[100]

WPO can be domestically sourced and stockpiled without relying on international infrastructure investment. Additionally, with proper policy backing and quality control, these fuels can contribute to sustainable energy targets and emissions reduction strategies. Depending on the refinery, cost savings for government fleet fuel can be achieved with these recycled fuels.

WPO is a mitigation strategy in response to the global recycling crisis, creating domestic markets for waste plastic.

Additionally, it can be a solution for hybrid energy systems in remote communities. WPO can replace traditional diesel without changes to the hybrid systems.

Quality control is essential to the success of a WPO refinery. In order to satisfy the scrutiny of global NGOs and governments, emissions need to be strictly tested and controlled.[101] The burning of plastics to create fuel will cause emissions, however, through quality controlled refining processes; those emissions might be less than traditional fossil fuels. Further research from a chemical engineering perspective is needed in this regard as this is an emerging technology and the emissions have not yet been quantified.

Therefore, support from the renewable energy sector will require WPO to have low emissions, high efficiency and to be refined using strictly solid waste plastics that are produced as an unintentional and unavoidable consequence of an industrial process. This ensures that the WPO is a recycled carbon fuel.[102]

The European Commission has passed directives along those lines to promote recycled carbon fuels as a separate category that, while not a renewable fuel, can provide the minimum greenhouse gas emissions savings to be included in the decarbonisation of the transport industry.[103] The Directive requires member states to set a minimum obligation on fuel suppliers to ensure that the share of renewable energy within the final consumption of energy in the transport sector is at least 14%. Recycled carbon fuels, if they meet the emissions reductions benchmarks, are allowed to contribute to that 14% target.[104]

If WPO meets the same *emissions reduction targets and quality control benchmarks* as biodiesel, it could be argued that WPO be given the same status as biodiesel.

Diesel can also be made more eco-friendly by adding an exhaust fluid made from urea, commonly known as AdBlue. However, a supply chain crisis that occurred when Chinese exports from China to Australia were halted drew attention to the sovereign risk of relying on imports for energy security.[105] Australia has enough local wind and solar to power the trucking industry if it converted to EVs. This would negate the sovereign risk in the current supply chain.

Summary

The case study selection encompasses renewable energy projects across wind, hydro, solar and biogas technology as identified and reviewed above. Future development projects should not be limited to these selections. As technology improves and innovation occurs, the small-scale technology solutions will also improve and change. The increasing affordability, efficiency and availability of the technology identified in this chapter can lead to more investment in these technologies for remote communities as will be explored in this industry guide. These innovations can be used to attract investment in sustainable renewable infrastructure for communities.

That case study analysis below is used to develop the set of Sustainable Community Investment Indicators that are based on the Sustainable Development Proposition (SDP) which is developed in this chapter.

The SDP can be used to attract investment in sustainable renewable infrastructure for communities and has been developed above as follows:

1. Environmental sustainability:
   a. Environmental sustainability can be upheld by following the precautionary principle and the principle of intergenerational equity.
   b. Valuation of natural capital resources, both renewable and non-renewable, has been analysed above and a novel approach to the natural capital aggregate rule has been developed. This can help to keep account of the renewable natural capital available for future generations.
2. Sustainable governance:
   a. A democratic justice system that affords liberty and human rights that supports a sustainable community has been reviewed. Property rights and economic justice that supports energy reliability and security is an important part of a democratic justice system.
   b. Legal governance structures and democratic community governance in energy projects leads to social benefits.
3. Sustainable technology:
   a. It has been shown that technology that increases yields from natural resources supports population growth while maintaining natural capital and improving standards of living.
   b. A review of renewable energy technology available for non-network solutions has been conducted.
4. Economic sustainability:
   a. The relationship among economic growth, increasing standards of living and supporting population growth sustainably has been outlined.
   b. A background of the energy marketplace is also explored.
   c. In the case studies below, economic sustainability has been found to favour systems that are economically self-sufficient and independent of reliance on outside ongoing public funding for maintenance and upgrades/decommissioning.
   d. The case studies will demonstrate that freedom of choice in financial and legal enterprise design is crucial to the economic success of a project and the ability to attract investment and raise funds.

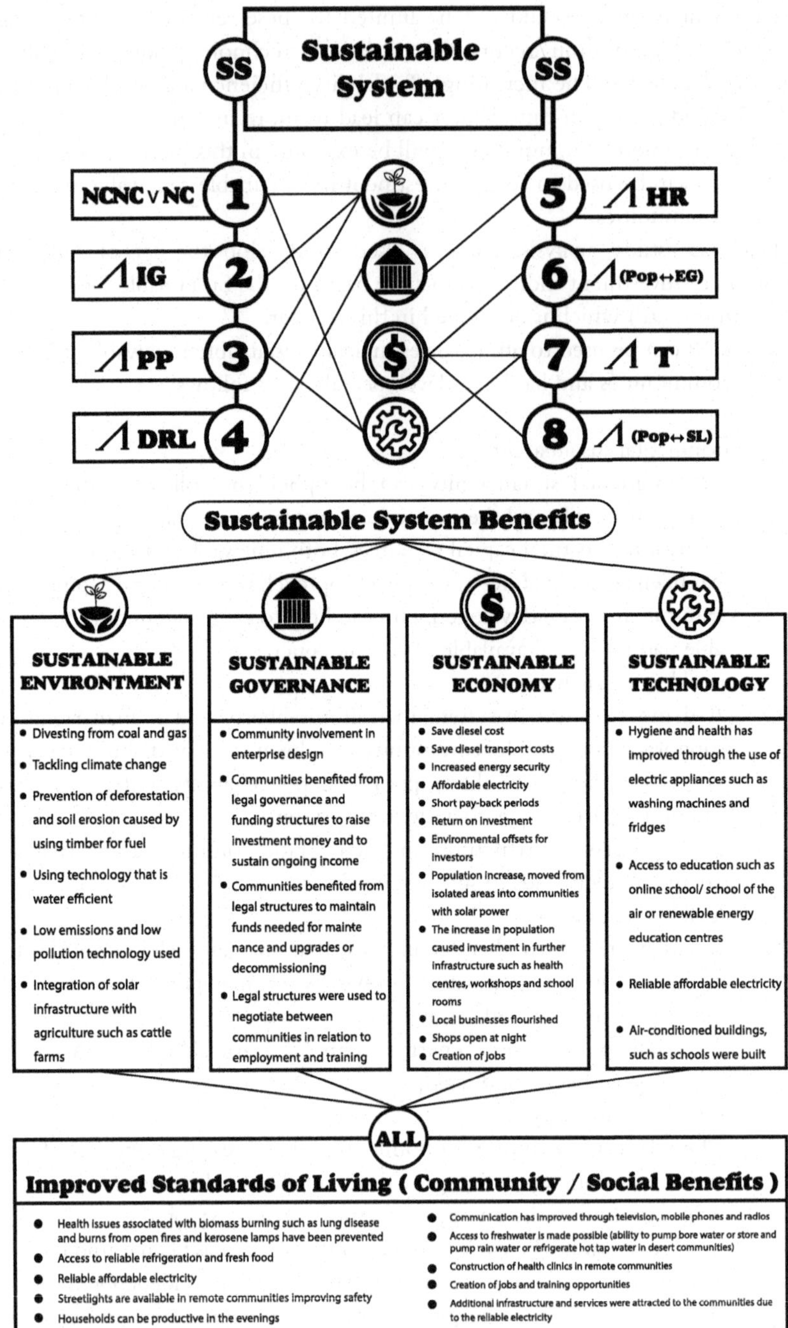

FIGURE 5.5 Diagram showing how the Sustainable Development Propositions relate to a sustainable system.

5. Sustainable system/community:
   a. When all of the propositions are satisfied, a Sustainable System or sustainable community exists.
   b. An increased standard of living that occurs due to improved environmental, technological, economic sustainability and improved governance means that social benefits are created such as improved access to electricity, health, education and employment.

Accordingly, it has been found that:

IF natural capital is enhanced, substituted, adapted OR remains constant AND intergenerational equity remains constant AND the precautionary principle is satisfied AND the democratic rule of law is upheld AND human rights and liberties are upheld AND economic growth increases in line with the population AND technology improves AND the standard of living increases in line with population growth THEN it is a Sustainable System.

The end goal of the SDP is that it can be used to develop systems that are sustainable and attractive for investment. Those systems should enable the trajectory of increased standards of living and economic growth for communities through improvement in technology while at the same time preserving the value of natural capital for future generations.

The Sustainable Development Proposition

Figure 5.5 shows how the SDPs relate to a Sustainable System. It shows that a Sustainable System or a sustainable community depends on all nine propositions and that some of those propositions relate to more than one category. For example, for technology to be sustainable, it must also be environmentally sustainable. The diagram also shows the SCIIs™ that can be achieved in a renewable energy project. Those SCIIs™ are benefits or outcomes of the Sustainable System.

Notes

1 Finkel 2017, 72.
2 Droege 2009.
3 Finkel 2017.
4 Ibid.
5 Ibid. 17.
6 Girardet 2008, 188.
7 Droege 2009, 18.
8 Star of the South; Preiss, B.
9 Coalition for Community Energy 2018. Webinar 7.
10 Ibid.
11 Homebiogas 2020.
12 Australian Renewable Energy Agency 2019. "Biogas Opportunities for Australia".
13 Sovacool et al. 2014. *Energy security, equality and justice*, 98.

13 Business Queensland 2016.
14 Business Queensland 2016.
15 Ibid.
16 Ibid.
17 Sovacool 2012. *Energy & Environmental Science*, 9157–9162.
18 Beasley industries Adelaide at www.beasley.com.au from Girardet 2008, 183.
19 Beyond Zero Emissions, The Energy Freedom Home, 2015, 51.
20 Xiaojing and Mackenzie.
21 Ibid.
22 Ibid.
23 Ibid.
24 Ibid.
25 The Australian Energy Regulator, Annual Report, 2020, 31; CSIRO, GenCost 2019–2020.
26 Pernick and Wilder 2014.
27 Australian Renewable Energy Agency, "Solar power in Australia".
28 Beyond Zero Emissions, The Energy Freedom Home, 2015, 51.
29 Finkel 2017
30 Flavin, Chris, Pres, World Watch Institute, interview for the People's planet TV series, from Girardet 2008, 186.
31 Finkel 2017, 74.
32 Ibid.
33 Mazengarb 2021.
34 Ibid.
35 Finkel and Graves. *The Future of Renewables.*
36 Fraunhofer ISE 2020.
37 Nguyen et al. 2017.
38 Ibid.
39 Ibid.
40 Ibid. 4056.
41 Li et al. 2018.
42 Myers et al. 2018, NASA, 1–4.
43 COAG 2013. "Revised National Principles for Feed-In-Tariff Arrangements".
44 Beyond Zero Emissions, The Energy Freedom Home, 2015, 80.
45 Institute for sustainable futures, 2016.
46 Droege 2009, 15.
47 Ibid.
48 Ibid. 15–16.
49 Beyond Zero Emissions, The Energy Freedom Home, 2015, 76.
50 295.Victoria Government, Department of Environment, Land, Water and Planning 2017. "Certificate of electrical safety."
51 *Electricity and Other Legislation (Batteries and Premium Feed-in Tariff) Amendment Bill* 2017, Explanatory Memorandum at pages 7–8. See also, The Construction and operation of solar farms Code of Practice (PN12493).
52 Victoria Government, Department of Environment, Land, Water and Planning 2017. "Certificate of electrical safety", 75.
53 Bird et al. 2022.
54 CSIRO manufacturing, energy, land and water. April 2018.
55 The Australian Institute: Seminar 20 August 2020.

56 Yang et al. 2015.
57 Ibid.
58 Ibid.
59 Vorrath 2016.
60 National Aeronautics and Space Administration, "An Overview of the Photovoltaic and Electrochemical Systems Branch at the NASA Glenn Research Center".
61 Beyond Zero Emissions and the University of Melbourne, *The Australian Sustainable Energy Zero Carbon Australia Stationary Energy Plan*, 2010, 24.
62 Way 2008.
63 Vorrath 2015.
64 Barber 2021.
65 Finkel 2017.
66 Ibid. See also Maisch 2020.
67 Redback Technologies.
68 AGL. n.d. 2020.
69 See, for example, Sonnen at https://sonnen.com.au/ and Reposit at https://repositpower.com/
70 Energy Queensland 2010.
71 The Australian Consumer Law applies to market power situations in VPPs in the same way as with peak demand supply market power situations on standard electricity grids. In addition, ring-fencing guidelines for DER are in place to prevent monopolies. See, for example, Australian Energy Regulator Annual Report 2020 at p.19.
72 The Australian Energy Regulator. Annual Report, 2020, 43.
73 Beyond Zero Emissions and MRC agency 2016.
74 Foley & Lardner Group 2020.
75 Rystad Energy.
76 Foley & Lardner Group 2020.
77 Ibid.
78 Beyond Zero Emissions and MRC agency 2016.
79 Ibid.
80 Horackzec 2021.
81 Lewis 2018, 47.
82 The Australian Energy Regulator, Annual Report, July 2020.
83 Ibid.
84 *The Treasury Laws Amendment (Consumer Data Right) Act* 2019.
85 The Australian Privacy Principles from Schedule 1 of the Privacy Amendment (Enhancing Privacy Protection) Act 2012.
86 Van Eygen et al. 2018.
87 Pickin and Trinh 2019. Australian Federal Government, Data on exports of Australian wastes 2018–2019 (version 2), 236.
88 Ibid.
89 Girardet 2008, 123.
90 Ibid.
91 McDonough et al. 2002.
92 Quality Circular Polymers 2017.
93 European Union, *Circular Plastics Alliance Declaration*, 20 September 2019.
94 Centre for Appropriate Technology Limited 2016, 62.
95 Damodharan et al. 2019.
96 Ibid.

97 Centre for Appropriate Technology Limited 2016, 62.
98 Pernick and Wilder 2014.
99 Ibid.
100 European Commission, Renewable Energy Directive.
101 Recycling Magazine 2020.
102 European Commission, Renewable Energy Directive.
103

> 'recycled carbon fuels' means liquid and gaseous fuels that are produced from liquid or solid waste streams of non-renewable origin which are not suitable for material recovery in accordance with Article 4 of Directive 2008/98/EC, or from waste processing gas and exhaust gas of non-renewable origin which are produced as an unavoidable and unintentional consequence of the production process in industrial installation.
>
> *from European Commission, Renewable Energy Directive (Article 2, 35)*

104 Ibid (Article 25).
105 Evans 2021.

6

CASE STUDIES

Introduction

Renewable energy infrastructure case studies can fall into the following funding categories:

1. Public – usually funded through tax base funded grants
2. Private
   a. Donation/non-profit
   b. Community investment
   c. Community-developer partnership
   d. Other private investment
3. Public–private partnerships

Each case study will comprise of a cost-benefit analysis on the sustainability of the project.

Case study selection

The case study selection is according to the following selection criteria:

1. The community case study:
   a. Was in a remote rural and/or Indigenous community;
   b. Did not initially have adequate energy supply (was off-grid or had a reliability or affordability issue);
   c. Used sustainable renewable energy resources in the infrastructure project;
   d. Was successful and provided an important learning element for future projects; and

DOI: 10.4324/9781003324669-6

   e. Involved an innovative or unusual financial, technological and/or legal structure;
2. The selection represented points of difference to other case studies across:
   a. Funding models;
   b. Legal and governance structures;
   c. Energy infrastructure technology;
   d. Political structures, that is, private, community owned, non-profit, government funded and donation funded; and
   e. A variety of international jurisdictions.

The criteria ensured wide ranging comparison of findings that could be triangulated and evaluated in a cross-case-study analysis to identify innovative strategies giving the best possible chance that a new innovative structure could be developed by combining successful elements from multiple sources.

Indigenous communities have been the subject of an enormous amount of intrusive surveying and research, particularly in instances where the research has not led to clear positive outcomes for the communities that have been surveyed. This can have a negative impact on the community as it has the effect of drawing attention to the fact that external researchers are aware of the issues the community faces but can offer no support. While many surveys and interviews of this kind have been extremely important in identifying issues that communities face, the time for identifying energy poverty issues has passed.[1]

In particular, researchers have lived with Indigenous communities and conducted interviews with those communities over extended periods of time in relation to energy security and standards of living.[2] Those researchers have interviewed indigenous communities extensively about the issues of power outages, power cards and power affordability and how this relates to daily lives of being unable to do laundry, refrigerate food or sleep inside where it is safe because of the lack of air conditioning.[3] If further interviews are conducted about these issues without solutions being brought, this would be insensitive to the standard of living gap that those communities face on a daily basis as it has already been well-established that a standard of living gap exists.

The purpose of this guide is not to re-identify those issues but to find solutions that can subsequently result in positive outcomes for those communities. Once those solutions are developed and shared, investors can use community consultation when it is appropriate and useful, at the pre-construction phase. This is when the investment is being sourced but before the project construction phase commences. For this reason, case studies were selected where the information was available in the public domain and further interviews were not necessary.

The aim of the research is to empower communities with knowledge of legal and financial innovation that can attract investment from non-indigenous private and public institutions. For this reason, the Sustainable Community Investment Indicators (SCIIs™) have been developed out of the case study findings.

The SCIIs™ are aimed at empowering communities to determine for themselves what values they will give weight to when seeking investors. Different communities will have the flexibility to give weight to different investment indicators depending on the direction that they want to take in their enterprise design.

Energy research is disproportionately focused on advanced energy technologies, commercial fuels and large-scale centralised power plants even though billions of people rely on wood, charcoal and other biomass fuels.[4] Such "traditional" fuels comprise 40–60% of total energy consumption for most communities and household cooking is responsible for 60% of total energy use in developing areas in Africa and can exceed 80% in some areas.[5] One quarter of household labour is often devoted to collecting fuel and the life endangering pollution that results from inefficient combustion is estimated to cause 4.3 million deaths a year due to chronic illnesses resulting from "traditional" fuels.[6]

Ninety-four per cent of Australian Aboriginal households in Northern Territory homelands and outstations still use firewood for heating and cooking.[7] This is considered by some elders, for example, in the Yolnu community, to be an important part of cultural heritage. Although elders are concerned with the health and safety aspects of open fire places, they often prefer cooking outside, sitting outside and sleeping outside next to the fire as this is in line with traditional ways of living. It has been suggested by researchers in the Yolnu community that if housing was designed to suit the traditional ways of life and designed to be appropriate to the climate of the community, energy consumption through air conditioning and heating would be significantly decreased.[8] In Figure 6.1 an African woman is using traditional fuels for cooking but is using solar power to charge her phone. This cultural preference is one example of how case studies can better inform than statistical

FIGURE 6.1 A young African woman is cooking corn maze, a staple food, on a wood fire stove under a hut. A solar panel is being used to charge technological devices. Photo by GCShutter from iStock Photo, Getty Images.

research methods as they add a depth of perspective to the individual communities that are being researched. The case studies demonstrate the importance of recognising the diverse cultural and economic needs of each individual community when conducting an analysis. Community research must be context based and account for the diverse communities rather than just treating community members as "numbers" or "statistics" to be measured.

Energy use, especially in developing communities, is gendered. For instance, women comprise the majority of those vulnerable to energy security. In African communities, the time spent in fuel collection can range from 1 to 5 hours per day, frequently with infants strapped to their back,

> the energy poverty nexus has a distinct gender bias .... Over the course of a typical year in Tanzania, a woman will spend almost 2000h dealing with chores, collecting wood, cooking, and other tasks compared with only 500h for men; that same woman will carry almost 90 tons whereas the same man will now carry less than 12 tons..... Just how hazardous is this to the health of mothers and children....[9]

Therefore, gender is an important perspective to include in such energy justice research. More female researchers are needed as less than 16% authors across a sample of 4,444 peer-reviewed articles were women.[10]

Case studies are seen as an underutilised human centred method of research into energy systems. Statistics on costs and prices or demographics do not take into account social and cultural factors, energy demands, and other interdisciplinary circumstances.[11] Comparative case studies "can more rigorously generate and test hypotheses across multiple areas, resulting in stronger evidence of convergence of findings."[12]

Table 6.1 provides a list of the case studies analysed in this chapter. The case studies are organised in the table by region, technology and funding structure.

Historical case study

Financial and legal innovation, often born out of necessity has been the key driver in the construction of energy infrastructure for many communities throughout history.

The following historical case study opens the historical analysis of such innovation. This case study is an example of a historical, non-network solution that emerged through financial and legal innovation. The historical case study demonstrates that free-market forces and legal and financial technology can overcome governance barriers to renewable energy solutions. In this situation, the innovative structure of the world's first joint stock corporation allowed the continuation of the wind energy infrastructure to continue for 700 years, demonstrating long-term sustainability.

TABLE 6.1 Case Studies

Name	*Organisation*	*Region*	*Technology*	*Funding Structure*
Historical				
Toulouse Water Mill	The Honor del Bazacle	Toulouse, south-western France, 1372	Hydropower (a shore based water mill) was built.	The world's first recorded joint-stock corporation since Ancient Rome raised investment through a share issuance of limited liability shares.
Remote Australian Indigenous				
Mabunji Aboriginal Resource Centre	Centre for Appropriate Technology	Borroloola, Northern Territory, Australia	Solar	Initial grant followed by an energy maintenance trust fund using income from housing rent.
Milibundurra Island	Centre for Appropriate Technology	Northern Territory, Australia	Solar	Grant funded, Bushlight programme.
Tafe At Mingalkala	Centre for Appropriate Technology	Gooniyandi Country, Western Australia	Solar	Grant funded Bushlight programme.
Gurrumurru	Centre for Appropriate Technology	Northern Territory, Australia	Solar	Grant funded Bushlight programme.
Kakadu	Centre for Appropriate Technology	Kakadu National Park, Northern Territory, Australia	Solar	Grant funded Bushlight programme working in partnership with the Federal Department of Families, Housing, Community Services and Indigenous Affairs.
Ulpanyali	Centre for Appropriate Technology	Uluru, Northern Territory Australia	Solar	A portion of tourism fees is paid into a trust fund. The trust fund money is used to build solar infrastructure as well as expand local businesses and facilities. Bushlight grants also contributed.

(continued)

TABLE 6.1 Cont.

Name	*Organisation*	*Region*	*Technology*	*Funding Structure*
Chile Creek	Centre for Appropriate Technology	Broome, Western Australia	Solar	Grant funded Bushlight programme.
Blue Bush	Centre for Appropriate Technology	Tennant Creek, Northern Territory, Australia	Solar	Grant funded through the Bushlight programme in partnership with the Julalikari Council Aboriginal Corporation (JCAC) via the Australian Government's Community Employment Projects (CDEP).
Desert Knowledge Centre (DKC)	Centre for Appropriate Technology in joint venture with Desert Knowledge Australia (a statutory corporation of the Northern Territory). Desert Knowledge Australia is a partner of the Desert Peoples Centre. The Desert Peoples Centre is a joint venture between the Centre for Appropriate Technology and the Batchelor Institute of Indigenous Tertiary Education. Ekistica Pty Ltd is a subsidiary company of the Centre for Appropriate Technology and is an engineering company that carries out the work. The Centre for Appropriate Technology has a joint venture with various Aboriginal corporations and private employment agencies (such as MyPathway).	South of Alice Springs.	Solar	Grant based with income returns following initial grants. Grants are received from various domestic and International governments and organisations including the Australian Renewable Energy Agency, The Northern Territory government and the University of Singapore.

Pacific Islander Communities				
Pacific Renewable Energy Project	A partnership between Austrade and the World Bank.	Twelve countries in the Pacific Islands: Palau, Federated States of Micronesia, Marshall Islands, Kiribati, Samoa, Timor-Leste, Papua New Guinea, Solomon Islands, Vanuatu, Tuvalu, Fiji, Tonga.	Small scale solar, hydropower, hybrid wind and solar, biodiesel.	Grant based. Currently trying to attract private investment.
Biogas				
Grameen Shakti	Grameen Bank	Bangladesh, India.	Solar and biogas	Microcredit and micro financing.
Biogas Projects	Grameen Shakti, China's REDP, Mongolia's REAP, Nepal's REDP, and Sri Lanka's EDSP	China, Mongolia, Nepal, Sri Lanka.	Biogas	Microcredit, low interest loans and donations/grants.
Rural Wind Energy				
Hepburn Wind and Samsø	A comparison between two wind energy projects, the first, Hepburn Wind, is an Australian project built in the rural community, Daylesford, Victoria. The second is in Denmark on Samsø, the first island in Europe to achieve 100% of its electricity from renewable energy, predominantly wind including EVs powered by wind energy.	Rural Victoria, Australia and rural Denmark.	Wind	The capital to fund the Hepburn Wind project initially came from a combination of issuing shares, loans and grants. Eleven small turbines (1 MW) were erected on Samsø Island at the cost of 8.8 million euros. Shares in the company were sold to the community members who received dividends from the sale of the power. Forty private applications for wind turbines were submitted so not all projects were approved. Some offshore wind generators were also funded in later years by the Danish Energy Agency at the cost of 33.3 million euros.

(continued)

TABLE 6.1 Cont.

Name	*Organisation*	*Region*	*Technology*	*Funding Structure*
Australian Rural Community Solar				
Clearsky Solar Investments	Clearsky Solar Investments	Rural communities in rural New South Wales (NSW), Australia.	Solar	Partnerships between community investors and private investors.
CORENA	CORENA	NSW, Queensland, South Australia, Victoria, Tasmania and Western Australia	Solar	Revolving Funds
Repower One-Five	Repower	Shoalhaven, NSW, Australia	Solar	Private investment
Lismore Council	Lismore Council	Lismore, NSW, Australia	Solar	Council-community investment partnership
Micro Hydropower				
Upper Yarra Community Power	Upper Yarra Community Power	Warburton, Victoria, Australia	Hydro	Initial grant, volunteer workforce, subsequent unlisted public company with share issue to raise capital.
Network Solutions for Agricultural Communities				
Bomen Solar Farm	Spark Infrastructure Group, subsequently Pika Pty Ltd.	Wagga Wagga, New South Wales, Australia	Large scale solar on grid	Public company which was bought by a private company and delisted from the stock exchange.
Terrain Solar	Terrain Solar Pty Ltd	Rural Australia	Large scale solar on grid	Private company funded by PPAs. Partnerships with landowners.

Rural Italy				
Varese Ligure	Local government and power company ACAM	Varese Ligure, Italy	Wind, solar, biomass and small scale hydro make up 100% self-sufficiency.	Initial grant funded, sustained in the long term by a joint private energy company (ARE Liguria SPA), public energy company (ACAM) and local government partnership.
Rural Latin America				
Light At Home	Acciona.org	Mexico and Peru	Solar	Non-profit organisation funded by public grants. Customers pay half the costs of the solar PV. Customers can obtain funding through micro finance.
Rural Africa				
Namibia	Elephant Energy	Namibia, Africa	Solar	Non-profit organisation funded by grants and donations.

Toulouse water mill[13]

1. Jurisdiction/region

Figure 6.2 depicts the Garrone River, the site of the original Les Moulins du Bazacle, a Watermill in Toulouse, south-western France. Toulouse began as a Celtic settlement, and then flourished as a Roman city before becoming the capital of the Visigoth kingdom. In the eleventh and twelfth centuries, Toulouse was ruled by the counts of Toulouse as an independent heritable fiefdom and it encompassed half of southern France and parts of Spain. The counts gave up political control over time through a social contract that involved military support from citizens in return for self-governance. Subsequently, after the year 1000, Toulouse was trade orientated with a strong merchant class and a tendency towards citizen rule and openness to religious freedom.

The law was different from other Mediterranean trading ports. Roman law was preserved by King Alaric II who commissioned codification of the Roman law in 506 CE and applied the code throughout the Visigoth Kingdom. This was developed over time into a flexible system of contracts, financial instruments and

FIGURE 6.2 A view of the Garrone River in Toulouse, France. Photo by Clovis Wood Photography from Unsplash.

property rights. Under this flexible system of governance and financial rights the local businesses flourished.

This case study takes place in Toulouse in the year 1372 when the Honor del Bazacle, Europe's oldest corporation, was formally incorporated in order to fund and govern the local wind powered mills.

2. Background of community participants

Leadership was held by the town council as a corporate entity responsible to the citizens. The Count was the ruler.

3. Technology used

Hydro power (a shore based water mill) was built. Figure 6.3 depicts an example of a traditional water mill.

4. Funding model

Money was raised by millers who pooled their capital and divided the mill into shares (one-eighth each). The profit was generated by the grinding of grain and ancillary rights such as fishing. The shares could be bought and sold to anyone. The dividends were paid in the form of grain at harvest time.

FIGURE 6.3 An example of a traditional water mill in a rural setting. Photo by Pixabay from Pexels.

5. Legal/governance model

The Honor del Bazacle was the first recorded joint-stock corporation since Ancient Rome. It is an early example of a company merger.

This corporate structure emerged in a fledgling capitalist climate based on an innovative flexible code of law adapted from Justinian's Code (Roman law) allowing a modern system of contracts and property rights well-suited to business and commerce. In particular, freedom for all citizens to contract financially through loans, mortgages, leases, sub-leases, fiefs, bailment and asset partnerships was encouraged.

The corporation was formed through the merging of 12 smaller mill companies that shared land near the Garonne River. This was made possible because of the property rights of those mills.

The company had a constitution, annual shareholders meetings, detailed financial record keeping and fully transferrable limited liability shares. The board of directors oversaw the managers and the managers oversaw the employees. It was treated as a separate legal entity and could hold property and write contracts in its name.

It survived as a public corporation until 1946 when it was nationalised as part of the French public electricity company EDF.

Cost-benefit analysis

6. Benefits

The success of the Honor del Bazacle was due to a combination of technological innovation (the shore based water mill as opposed to the traditional barge water mill), property economics innovation (the merging of land owners property rights), legal innovation (the formation of a company) and financial innovation (the issuing of shares to fund raise).

This innovation was developed in response to the inadequacies of floating mills. The floating barges which carried the early paddle-powered mills were overcrowding the river and causing crashes and struggles for the best positions on the river. The building of a grounded mill was a solution but required owning or leasing the property, additionally it required significant funding to build the infrastructure (the mill).[14]

The company provided a steady source of income to citizens in the middle ages, a period of serfdom.

7. Costs

Planning approvals for mills were difficult to obtain and dependent on the count. In 1138, four partners acquired the right to develop the mill by means of transfer of fief from the count. One of the partners was a prior; this helped with the planning approval because influence was often a barrier to development approval. The investors pooled their capital, acquired the development rights from the count and built the mills. They then merged with eight other milling companies.[15]

The limited liability of the company enabled it to endure natural disasters such as floods and fires. In those cases where a shareholder could not afford to contribute money to rebuild, they could surrender their shares and not have pay the unlimited amount, thereby putting a limit on the downside risk the investor faced. This encourages investment in risky infrastructure projects.[16]

One example of financial innovation was an enterprising hydrologist who offered to rebuild the mill after a flood in return for shares. He sold some of the shares to finance the build and kept the other shares to return a profit after the rebuild.[17]

The survival of the company for over 700 years is attributed to the specific contract, corporate, financial and property rights protected by the democratic laws of the society.

8. Learnings and recommendations

The corporation survived the 100 Years War, French Revolution, all of the changes in government and despite the infrastructure itself (the mill and buildings) being burnt or swept away by flood numerous times, the basic financial and legal structure survived and enabled the production of energy to continue.[18]

The combination of an innovative investment vehicle that brought together capital investment and property rights empowered the new economic class, who lived on their dividends of their investments. The company was created to be perpetual and in fact existed for around 700 years before being nationalised. The company survived in spite of competition from rival firms upstream.[19]

The innovation was made possible through the formation of private contracts and did not depend on royal decree or governance by the count.

The corporation was designed in great detail with a long company charter, such that it could exist autonomously for centuries. The company charter set out the rules such that no single participant could "cheat or ruin the game." The participation was voluntary. Participants had to volunteer to give up their property or milling rights in order to merge, they were not acquired compulsorily. This is an essential element of a democratic process of joint ownership and operation. The initial company charter is important because it ensures the continuing fairness for future generations of shareholders. The price of the shares reflected the trust the shareholders had in the company's autonomy and its continued protection of capital and fair return on capital.[20]

Overall, this case study demonstrates that legal and financial innovation can enable renewable, long-term sustainable energy infrastructure builds that benefit both communities and investors. Where there are current barriers to this, innovative and hybrid structures will be investigated and developed.

Australian Indigenous communities

Australia is in a unique position to provide case studies on rural and remote farming and Indigenous community energy infrastructure. Due to the vast distances between populations, many remote communities require stand-alone off-grid or micro grid, non-network and network solutions to energy supply. The other unique aspect of Australian remote communities is that the land in those communities is often subject to a wide range of property interests that are held over the same parcel of land. For instance, native title, pastoral lease, mining interests, telecommunications companies, government departments and local Aboriginal Land Council interests often co-exist.

Additionally, many remote Australian communities are still reliant on diesel generators. This means that if supply of diesel is cut-off by flood, bush fire or other delays, communities have power outages.[21]

Australia has the financial resources to improve the standard of energy supply infrastructure to those remote communities and has begun a process of public, community, non-profit and private investment in such projects.

Accordingly, Australia is a source of many useful community energy infrastructure case studies. The Australian case studies will be contrasted with international case studies to provide a depth and range of global energy infrastructure solutions for different kinds of rural environments and communities.

Current state of infrastructure in Australian Indigenous communities

In Australia, Bushlight ran a renewable energy programme from 2002 to 2013 and installed approximately 130 solar systems during this time period across Queensland, the Northern Territory and Western Australia.[22] Additionally, Bushtel lists 35 sites where solar infrastructure has been installed in Indigenous communities in the Northern Territory in recent years.[23]

Australian State, Federal and Territory governments have pledged significant funding to build solar infrastructure in remote communities in the past two years. The Kimberly Region received Federal government funding for 50 small solar projects in Indigenous communities as part of the Remote Indigenous Energy Program.[24] In Queensland, several solar energy projects have been funded by the State government as part of a new Decarbonising Remote Communities program. At the end of November 2019, in the Indigenous community of Doomagee, a 304 kW ground-mounted solar farm extension (adjacent to the existing 264 kW solar farm) and 105 kW of rooftop solar was installed on Doomadgee Shire Council roofs. The Doomagee Council will save approximately $30,000 a year on electricity bills. Fifteen construction jobs were provided to the local community. Similar projects were commissioned for Mapoon and Pormpuraaw, with the latter's solar infrastructure expected to save around $40,000 for the Indigenous Council annually over the next 20 years. On the Northern Peninsula, 340 kW of rooftop solar is planned for installation across the Bamaga, Umagico, Injinoo, Seisia and

New Mapoon communities. The NPA Regional Council will save approximately $87,000 per year on electricity.[25]

The Australian Federal government also pledged more than $19 million to 17 micro-grid projects in 2020 through the Regional and Remote Communities Reliability Fund. It funds Indigenous communities' renewable energy projects in the Northern Territory, Western Australia and Queensland as well as biofuel for a dairy farm in New South Wales and renewable hydrogen production. The goal is to provide reliable, low-cost, off-grid power supply and to reduce remote communities' reliance on expensive diesel generation. Part of the goal is to increase employment in regional communities.[26] The Microgrids programme will run over five years to 2024.[27]

The Northern Territory government has partnered with the Federal government through the Australian Renewable Energy Agency (ARENA) to access the Remote Communities Reliability Fund through this five year programme. The Northern Territory will install 25 off-grid solar systems for remote Indigenous communities.[28]

There are other small solar projects that can be identified, for example, the Queensland University of Technology (QUT) was involved in a funded solar project with the Kooma Traditional Owners Association.[29]

The Queensland Government has also identified future infrastructure supply to remote communities in the latest annual report, noting that Energy Queensland is focused on reducing reliance on diesel in remote communities.[30] In particular, Energy Queensland is trialling a 200 kW rooftop solar farm on government buildings in the Lockhart River region. A 16.9 kWh battery energy storage system helps supply around 10% of the communities' electricity.[31] Energy Queensland estimates that 1.5 MW of rooftop solar and 2,300 MWh of renewable energy are powering isolated regions.[32]

Aboriginal communities, also known as homelands or outstations, can have between 50 and 150 residents. Most of the funding for infrastructure comes from government. Resource agencies provide the services to the communities.[33] The latest surveys, reported in 2016, found that many communities surveyed at the top end of Australia have no energy supply or rely on small diesel or petrol generators. Most of the connected residents pay for their own electricity through power cards.[34]

> A quarter were reliant on small diesel or petrol generators for their energy service, which subjects them to ongoing cost pressures resulting from fuel price increases. This raises the potential to re-visit the provision of hybrid (diesel, battery, PV) systems for isolated communities. The suitability of hybrid systems as an energy supply in remote areas is well documented. While the capital costs of hybrid systems are higher, lifecycle costs (assuming systems are maintained correctly) are lower, potentially reducing the cost burden to residents and governments, as well as reducing CO_2 emissions. Technologies are advancing rapidly in this field, with a new generation of batteries and PV cells entering the market, and costs are decreasing.[35]

FIGURE 6.4 Three technicians, including two Indigenous Australians, walk past a remote solar installation in regional Australia. Photo by Thurtell from iStock Photo, Getty Images.

Case studies where these technologies have been trialled in remote communities are analysed below.

Australian Indigenous case studies

CfAT (Centre for Appropriate Technology)

1. Jurisdiction/region

Remote communities at the top end of Australia, Western Australia, Northern Territory and Queensland.[36]

2. Background of community participants

The Centre for Appropriate Technology Limited (CfAT Ltd) was initially established to work with Aboriginal and Torres Strait Islander communities in the top end of Australia.[37] It is based in Alice Springs.[38] Its Bushlight programme was a government funded initiative.

There are numerous barriers to energy infrastructure investment in remote communities.

One of the primary reasons that energy infrastructure projects fail is property ownership issues are not dealt with adequately in the early stages of planning.[39]

Most infrastructure projects require complicated property negotiations at the outset, whether the project is remote or urban. This hurdle is exacerbated in regional areas where security of tenure issues must be overcome. For example, it is difficult to obtain a loan using native title interests, Deed of Grant in Trust (DOGIT) or pastoral lease interests as opposed to freehold ownership because a loan usually requires an income in order to service the loan or an alienable asset as security.[40]

Native Title Representative Bodies (NTRBs) are not always consistently recognised as representing the interests in any given area when unregistered and disputes have been known to arise.[41]

This uncertainty of tenure arises even where a native title determination has been made. A Registered Native Title Body (RNTB) is recognised by the Federal Court and an exclusive native title area has been determined, as this determination does not necessarily mean that an Indigenous community has freehold tenure. Without freehold tenure, there is no security to put up in order to secure a loan to build infrastructure. Additionally, recognition at law of customary economic rights has failed and a right to trade in resources has been rejected.[42]

Land is therefore often managed through ILUAs (Indigenous Land Use Agreements). ILUAs are a common way for energy companies to build infrastructure in a region, for example, the ILUA between Hazel Windsor and Colin Saltmere on behalf of the Indjalandji-Dhidhanu People and Indjalandji-Dhidhanu Aboriginal Corporation (ICN 7791) and Ergon Energy Corporation Limited dated November 26, 2012, as recognised in the native title consent determination, *Saltmere on behalf of the Indjalandji-Dhidhanu People v State of Queensland* [2012] FCA 1423.

However, where a determination of native title has not yet been made over a region, the right to negotiate an ILUA becomes uncertain, see for example, *Kemppi v Adani Mining Pty Ltd (No 4)* [2018] FCA 1245. Additionally, if a native title group does not want to accept an ILUA, the company they are negotiating with can apply to the National Native Title Tribunal (NNTT).[43]

There are some instances where a native title determination has been made and exclusive possession native title areas have been granted and fee simple tenure has also been granted. For example, the Bidunggu Land Trust area in *Aplin on behalf of the Waanyi Peoples v State of Queensland* (No 3)[44] was recommended by the *Aboriginal Land Tribunal* to be granted in fee simple by the Minister to the Waanyi People. The Minister accepted the recommendations. This was the last claim to be heard by the *Aboriginal Land Tribunal*, there being a sunset clause in The *Aboriginal Land Act 1991* to receive and hear claims, which provided no new claims could be lodged after December 22, 2006.[45]

The Bidunggu Land Trust area was a DOGIT before it became freehold via the above legal process. Aboriginal community councils are responsible for their communities under the *Local Government Act 2009*. They are also, under the *Land Act 1994* the trustees of Aboriginal DOGIT land. In 2008, Aboriginal Shire Councils were amalgamated into Regional Councils.

> Aboriginal shire councils therefore have dual responsibilities—first, as a local government under the *Local Government Act 2009*, and second, as the trustee of Aboriginal DOGIT land under the *Land Act 1994* with leasing powers, under the *Aboriginal Land Act* 1991.[46]

In towns such as Cherbourg, Queensland, this means that the community members are perpetually leasing their housing from the Aboriginal Land Council and any infrastructure projects must be community projects rather than individual housing projects.

The Aboriginal shire councils, as trustees, have extensive obligations, similar to other local government entities and must comply with legislation when building infrastructure, for example, the *Land Act 1994*, *Aboriginal Land Act 1991*, *Native Title Act 1993 (Cth)*, The *Planning Act (Qld) (2016)*, s 5, *Local Government Act 2009*, *Vegetation Management Act 1999*, *Aboriginal Cultural Heritage Act 2003* and any other Acts that require statutory processes.[47]

One other issue that project developers must be aware of is that some native title claims over an area are unsettled and inconsistent with DOGITs, local government and State interests and pastoral leases or freehold land owners.[48] Additionally, the Aboriginal shire councils sometimes hold two roles as both trustee and also under the local government legislation when deciding on whether to grant leases for assessable developments. This might at times be in conflict and must be managed through transparent council meeting minutes.[49]

In summary, there are barriers to infrastructure builds in Indigenous communities and the innovative legal and financial solutions that are deployed in these communities will depend largely on the underlying tenure and property rights and interests.

3. Technology used

There are four types of micro-grid energy systems that CfAT reports as being used to deliver electricity to remote Indigenous communities.

a. Diesel/petrol generators (micro grids or individual systems);
b. Solar PV (micro grids or individual systems);
c. The grid; and
d. Hybrid (combination of solar power and backup diesel generators).[50]

Other than the grid supply, the most reliable energy supply was from the hybrid systems with 76% of power being supplied for 24 hours, the disconnected percentage often as a result of turning the generators off at night.[51]

Bushlight

Various Projects have been commissioned by CfAT including Bushlight, which deployed individual solar systems, hybrid systems and micro-grid systems to Aboriginal and Torres Strait Islander households.[52]

FIGURE 6.5 Australian Indigenous child playing. Photo by KerrieKerr from iStock Photo, Getty Images.

Bushlight has produced 35 case studies describing the success stories of this project. The case studies demonstrate that the communities with renewable energy systems have an improved standard of living with access to reliable refrigeration and fresh food, reduced cost of diesel fuel, ability to pump bore or rain water, access to school of the air and improved employment and training opportunities.

The successful case studies that demonstrated innovation are highlighted below:

Mabunji Aboriginal Resource Centre[53]

Bushlight uses its own solar systems, stating a 20 year lifespan for the panels and a ten year lifespan for the battery banks. Bushlight states that it is not capable of replacing the battery system once it is at the end of its lifespan. It states that the funding structure doesn't allow it. However, the Mabunji people established their energy maintenance trust fund in 2004 to prepare for the maintenance and replacement of the systems that Bushlight had supplied.[54]

The Mabunji People established the fund through the fortnightly housing rent payments of $65. Of that $65, $15 is deposited into the trust to be set aside for energy maintenance. Some of the residents were reluctant to allow the board to manage their money. However, residents have since accepted the structure. Significant work was done to reassure people that their money was properly managed. Some battery banks have already been replaced. Bushlight uses lead battery systems that require regular maintenance and top-ups of non-iodised water. Technology has since improved. The Mubunji Trust has been successful in replacing battery banks and also in paying for maintenance by Bushlight personnel.[55]

Milibundurra[56]

Before this Bushlight project commenced, the community was spending approximately $9,000 a year on diesel and $5,800 on diesel transport. When the diesel generator broke, Bushlight installed solar power on the two houses in the community. They also installed a water tank as the closest freshwater is 15 km away and is poor quality bore water. The Mubunji people subcontracted from Bushlight to install the systems. They received training. It is estimated that less than $1,000 will now be needed for diesel annually. A very positive outcome.[57]

Mingalkala[58]

The Bushlight system at the industry college has been upgraded several times as it has attracted an increase in students since the provision of reliable electricity.[59]

Nearby, local aboriginal businessman Stanley Till donated three demountable buildings to establish a course in construction for local Aboriginal people. He is seeking funding for a power supply to those buildings.[60]

Bushlight systems have been used by mothers to educate their children. It means they can get access to home-schooling resources. One mother has set up a small classroom with the Bushlight system that allows local aboriginal children to access school of the air via a satellite dish.[61]

Gurrumurru[62]

Bushlight systems were built to supply energy to five houses and the local school. Since solar power was introduced, the population has grown.[63]

Traditional Owner, Waralka, explained: "Everyone is coming back to look at the new Bushlight system."[64] With the installation of a low cost and reliable energy supply, life becomes a bit easier in remote areas and this often encourages more family to return to their homelands. In addition to this, there was the usual population fluctuation from cultural ceremonies and events. Population growth and infrastructural developments in Gurrumurru has required ongoing communication with residents to keep track of their changing energy needs. Bushlight, in collaboration with Gurrumurru and Laynhapuy, rolled out a series of solutions to support the changes as they occurred.[65]

Following this installation, a new health clinic was built and the Bushlight system was expanded as much of the existing infrastructure was reused and maintenance was shared. The community nurse could manage the energy via the already established Bushlight energy management unit (Bushlight personnel were paid to conduct maintenance). The system has since been expanded to support the energy needs of new residents coming to live in the town.[66]

Power outages have occurred due to the backup generator not functioning properly and have since been rectified. The system was finding it difficult to cope with the energy use of the community as it expanded but Bushlight is looking at solutions to these issues.[67]

FIGURE 6.6 Aerial view of Kakadu National Park, the Northern Territory, Australia. Photo by Vladimir Haltakov from Unsplash.

Despite these issues, the project has been a success,

> As Gurrumurru grows residents continue to think about how they can improve infrastructure and services to make the community a more comfortable place for everyone to live. Residents have expressed they are very happy to have solar power as they are saving money on diesel and can always rely on their fridges and freezers to keep their food fresh. Many improvements to infrastructure have already been made.[68]

Kakadu[69]

Bushlight installed a system in Kakadu National Park, an iconic tourist destination depicted in Figure 6.6. The system gives 24-hour reliable power supply and has reduced the annual cost of diesel from $150,000 to $20,000 across four homelands. They worked in partnership with the Federal Department of Families, Housing, Community Services and Indigenous Affairs. Four automated hybrid systems were installed. 160 tons of greenhouse gas emissions were saved annually. $40,000 in generator maintenance and services were also saved annually. Residents are still issued with power cards. This can mean power cuts if the card needs to be topped up.[70]

Ulpanyali[71]

This community has a trust fund for local development projects. A portion of the entry fee paid by tourists to enter Uluru (depicted in Figure 6.8) is paid into this trust fund.

Although the community is close to the grid, the logistics and cost of connecting a small community to the grid were deemed unviable. Therefore, a Bushlight renewable energy system was installed using the trust funds. The community states that they have saved money on diesel for the backup generator as well as the drive to Alice Springs for fresh food as they now have reliable fridges and freezers.

The generators previously used would cause the fridges and freezers to defrost when not running and the food would often be spoilt and wasted. The children are now able to drink more water, because it is refrigerated. Previously, the water was too hot to drink in summer because the pipes were so close to the desert surface. Since getting renewable energy, the community has established community farms, a tour business and a market garden. They have also used the trust money to build an art centre and a mechanical workshop. The local Indigenous women have developed an art business selling paintings at a local resort. The workshop is being used to undertake repairs and maintenance locally. Schoolchildren are now able to learn through school of the air while remaining on country and maintaining their culture and language. The community reports that there are now jobs available and they can support more people.[72]

Chile Creek[73]

Local community leader, Roma Puertollano, has led the construction of a small tourist resort at Chile Creek. Funded independently by the community, the resort was originally reliant on a diesel generator. The renewable energy Bushlight system has been installed and supplies 24-hour renewable energy to the four tourist huts and toilets. Since then, grants have been received to upgrade the tourist accommodation and Roma states that there are jobs for people in the community as she needs someone to help her run the tourist business and organise the community.[74]

Blue Bush[75]

Blue Bush is an outstation on an Aboriginal land trust near Tennant Creek. Figure 6.7 depicts the spectacular granite rock formation that is a tourist attraction known as the Devil's marbles. It is located near Tennant Creek in the Northern Territory, Australia.

The residents of Blue Bush are Bunnie Hooker, traditional owner and her husband, Norm Hooker, who grew up on a nearby cattle station. Prior to the renewable energy system, they spent approximately $20,000 a year on diesel to run the cattle farm. The 48 V community solar renewable energy system that was installed by Bushlight cost $211,000. Since installation, the diesel generator is rarely used. They have subsequently expanded their cattle farm and as agreed with, Bushlight employ locals from town to help with mustering and farming. They also accommodate local university students (Charles Darwin University) who visit to study the

FIGURE 6.7 Devil's Marbles, near Tennant Creek in the Northern Territory, Australia. Photo by Callum Parker from Unsplash.

farm. They manage their own maintenance. Bushlight trained the couple in system maintenance so that they only call Bushlight after first consulting the manual. The system itself was funded by and is owned by the resource agency, Julalikari Council Aboriginal Corporation (JCAC) via the Australian Government's Community Development Employment Projects (CDEP).[76]

Projects post Bushlight

Desert Knowledge Centre (DKC)

The Desert Knowledge Centre (DKC) is a federally funded solar city project aimed at connecting to the National Grid in the future.[77]

The DKC has a holistic approach to renewable energy, constructing education and training facilities and knowledge centres as well as renewable energy generation. The aim being to keep the local community engaged and trained in managing and building their own infrastructure.

- The Intalyalheme Centre for Future Energy is listed under the DKC annual report (funded by the Northern Territory Government with $5 million over three years). The goal is to power the NT with 50% renewable energy by 2030.[78]

Ekistica

Ekistica has delivered the following projects:

- Waterloo Wind farm
- Darwin International Airport 4 MW PV system + 1.5 MW expansion
- Cook Islands solar power
- Fiji micro projects

Analysis

1. Funding models

The funding model is grant based with some income from the infrastructure built, such as newly constructed solar centres, rent and other operations.[79]

In summary:

a. Bushlight

Grant based, trust funds and private investment.

b. CfAT

Grant based.

c. DKR

Grant based leading to income returns. For example, direct grants of over $1, 500,000 for the DKC alone were received in the last financial year. Additionally and indirectly, grants have been attracted that support this project. For example, Bushlight received $200,000 from the University of Singapore to build the solar infrastructure needed in the Desert Knowledge Precinct and Business Innovation Centre. Additionally, Ekistica, the subsidiary company of CfAT (the company responsible for Bushlight), was awarded funding to consult in relation to the off-grid power systems centre, Intyalheme.[80] The yearly income from this asset and the expenses for the DKC are listed in the 2019 Annual Report.[81]

d. Ekistica

As a subsidiary of CfAT, Ekistica paid a fully franked dividend of $120,000 to CfAT in the previous financial year.[82]

2. Legal/governance model

The underlying tenure for an Indigenous Renewable Energy project is crucial to its success. For example, with the DKC projects, the land is managed by Desert

Knowledge Australia (DKA). The land use is subject to an ILUA with the Arrente People. The Desert Peoples Centre is a joint venture project between the Batchelor Institute of Indigenous Tertiary Education (BIITE) and CfAT.[83]

There are many tenants on the 73 ha property which is south of Alice Springs, including government, not-for-profit (NFP) and other businesses.

Summary of the governance structures:

a. CfAT

CfAT is an incorporated NFP company.[84]

It is an Aboriginal and Torres Strait Islander controlled corporation formed under *The Corporations (Aboriginal and Torres Strait Islander) Act 2006 (Cth)* (CATSI Act).

CfAT works in partnership with government and non-government organisations and Aboriginal Land Councils.

b. DKA (joint venture with CfAT)

DKA is a partner organisation to the Desert Peoples Centre which is a joint venture between CfAT and BIITE.[85]

DKA is a statutory corporation of the Northern Territory. Working in collaboration with the ARENA, a memorandum of understanding has been established with this federally funded agency. The mutual aims are to support local non-network renewable energy systems that can be later connected to the Alice Springs Future Grid Project.[86]

c. Ekistica

CfAT's subsidiary company, Ekistica Pty Ltd is an engineering company.[87]

d. Other joint ventures

CfAT, as an NFP Aboriginal and Torres Strait Islander corporation, has a strategy to develop "separate commercial 'spin-off' companies to undertake fully commercial businesses, established over ten years ago with the launch of Ekistica Pty Ltd."[88] For example, they have established a 50:50 joint venture with MyPathway through a private company, CFATMPJV, the CfAT CEO Dr Steve Rogers and Chair Peter Reenchant were appointed Directors of CFATMPJV. Through this vehicle, they are entering into unincorporated joint ventures with local Aboriginal Corporations and are awarded Commonwealth Government funding to deliver projects to Aboriginal Communities.[89] The aim of these joint ventures is to maximise local and Aboriginal employment.[90]

Registered Native Title Bodies Corporates and Aboriginal Land Councils sometimes come into conflict, especially in cases where both organisations might be governed by the same people or in cases where native title claims are not determined or are determined differently to what was assumed by the community who set up the Aboriginal Land Corporation.[91] The RNTBCs are formed to manage native title interests under the *Native Title Act 1993*.

Aboriginal Land Councils are responsible under local government laws. Aboriginal and Torres Strait Islander corporations, such as CfAT are governed by the *(Corporations Aboriginal and Torres Strait Islander) Act 2006* (CATSI Act), as described below. Sometimes, as in the case of CfAT, they can enter into joint ventures and ILUAs with RNTBCs and Aboriginal Land Councils for the purposes of delivering community projects. Where an RNTBC has been determined by the Federal Court, they must register under the *CATSI Act*. So an RNTBC will also be an Aboriginal and Torres Strait Islander corporation but an Aboriginal and Torres Strait Islander corporation is not necessarily always an RNTBC.

Without a partnership with an Aboriginal and Torres Strait Islander corporation such as CfAT, RNTBCs can be limited in their funding arrangements, often having no source of income or assets that can be used as a security for a loan. Sometimes an RNTBC or an Aboriginal Land Council can be a trustee over former DOGIT land.

CfAT was awarded $1,115,509 last year in Commonwealth grants through its efforts as a registered Aboriginal and Torres Strait Islander corporation.[92]

e. Benefits

The benefit of an Aboriginal and Torres Strait Islander Corporate structure is the empowerment of Indigenous communities in their ability to negotiate with government and other private interests. Incorporation offers a legal entity under which a community can conduct business. It has been viewed as an "intercultural phenomena" because although corporate structures are not traditional Indigenous ways of engaging, they can be used by Indigenous communities to engage with the wider business community.[93]

Aboriginal and Torres Strait Islander corporations are said to provide a balance between the western concept of incorporation and the traditional Indigenous community governance as they allow internal governance rules. The system of replaceable rules in the *Commonwealth Corporations Act* is consistent with the CATSI Act and allows this "intercultural phenomena" to occur.[94]

The key difference to the *Corporations Act (Cth)* is that under the *Aboriginal and Torres Strait Islander Act*, the Registrar plays a significant role. The Registrar has powers to assist an Aboriginal and Torres Strait Islander corporation and to limit its actions. For example, the Registrar can prevent a corporation from altering its rules and has the power to change the constitution of the corporation if it is considered to be failing to meet its internal governance rules.

> A balance must be achieved between the application of Western and indigenous concepts to the running of indigenous corporations. It seems appropriate that this balance be struck so as to ensure that the internal management of indigenous corporations is based on culturally appropriate rules while the external accountability of indigenous corporations is governed by Western

concepts, since misconduct by indigenous corporations may have a negative impact on outsiders.[95]

The Registrar provides ongoing support to Aboriginal and Torres Strait Islander corporations. The Registrar provides compliance training, troubleshooting sessions and corporate governance training. The Registrar can offer dispute management services and can provide non-binding advice to corporations who are mediating. The Registrar can also provide advice and investigate complaints in relation to corporations. There are whistle blower provisions that the Registrar oversees. The Registrar also has the power to call general meetings if there is a concern. Additionally, the Registrar can call a meeting of interested people such as a funding body, creditors and other corporations to work out financial or governance problems. Small corporations must keep proper up-to-date financial records and report to the Registrar every second year. They also must have a general meeting every second year which can be done by video or teleconference. This helps reduce reporting costs. A corporation with an income over $100,000 needs to provide additional reporting. Large corporations with an income more than $5 million must provide reports consistent with the *Corporations Act 2001*.[96]

Clarity and training on internal governance and fiduciary duties has been recommended by commentators.[97] However, there are clear governance standards in the Act as it currently stands:

i. A civil penalties scheme;
ii. Disqualification measures;
iii. Directors and officers and managers duties including, a duty of care and honesty, duties of disclosure, a duty to avoid conflict of interest and a duty not to trade while insolvent;
iv. Members' participation including, members can request information about directors' payments and transactions involving another business or personal interest of a director or a relative of a director; and
v. Members' rights including, to inspect the books or to stop a corporation from acting in an unfair way.[98]

A key benefit to an Aboriginal and Torres Strait Islander corporation is that it provides a vehicle with which to attract investment or funding. An investor looking at funding an Aboriginal and Torres Strait Islander corporation is afforded the same protections as any other corporation, including external administration provisions of voluntary administration, receiver or liquidator provisions and applications to a court to seek an order to protect assets. Additionally, you cannot be a company or a company director if you've been declared bankrupt or disqualified from managing a corporation.[99]

The main difference between the *Corporations Act* and the *CATSI Act* is that the Corporations Act is governed by ASIC and the *CATSI Act* is governed by the Registrar who is supported by the Office of the Registrar of Indigenous

Corporations (ORIC). One advantage of the *CATSI Act* is that there is no limit on the number of members unlike the *Corporations Act* which has a limit of 50 members for a private company. They also do not have to pay lodgement fees to the Registrar. Additionally, the members will not be personally liable for the company's debts under the *CATSI Act*; however they are subject to civil penalties laws and they are can also be held personally responsible for actions related to the corporation under the duties of directors and other officers. Also, the members of an Aboriginal and Torres Strait Islander corporation cannot own or trade shares in the corporation. They cannot issue debentures or other securities. They are also prohibited from providing financial services or trade union services.[100]

Although technology requires tools, such as legislation and corporate structure and contracts, it also requires a cultural framework for analysis and critique.[101] The Aboriginal and Torres Strait Islander legislation is one such tool that can provide a cultural framework for analysis, critique and where necessary be amended.

f. Costs

Grant funding is usually an upfront amount of money, not an ongoing source of income. Consequently, there have been issues related to grant funded infrastructure programmes such as Bushlight. For instance, the program was funded by grants until 2013. Following this, the maintenance and upgrades of the solar systems in remote communities became the responsibility of the various local councils, individual outstations and Indigenous corporations. The remote locations coupled with lack of skilled workforce meant that many of the systems were not upgraded or maintained.[102]

Ekistica reported that grant funded energy supply projects in remote communities have had the following issues:

> High capital cost and difficulty accessing finance and funding; Poor reliability in remote locations; Lack of maintenance and limited access to experienced maintenance personnel; Lack of backup generation and support systems; Mismatch between demand for electricity and capacity to supply.[103]

A summary of the challenges identified with the Bushlight projects as found in 260 surveys included:

i. Maintenance, servicing and battery replacement needs to be accounted for in the cost analysis.
ii. Distance to the nearest service centre and availability of local trained people are relevant to the success of the project.
iii. Warranties must be enforced and honoured, which is not always the case.
iv. Initial capital costs and reliability are the biggest issues for farmers and Indigenous people.

v. Small power systems, less than 5 kW were more reliable and better received by communities.
vi. Transport of the systems to the remote locations was one of the most significant parts of the set-up costs.[104]

g. Learnings and recommendations

In response to those challenges listed above, the following recommendations are made:

i. Training for local maintenance and installation personnel in remote locations is needed, such as online TAFE or CfAT training programmes.
ii. Community engagement, such as the need/desire for renewable energy is crucial to the success of the project.[105]
iii. Innovative financing mechanisms are needed, instead of relying solely on initial grant funding.
iv. A structure and plan for ongoing income, such as a trust or Aboriginal and Torres Strait Islander corporation that generates income, is needed to sustain the project once the grant funding ends.
v. Reliable systems with warranties are crucial.
vi. Consumer protection enforcement for customers in remote communities is needed.

Additionally, in grant funded projects, Community Service Agreements (CSAs) should be well thought out and the parties held accountable. The CSA is an agreement between the community, its support or resource agency, the agency funding maintenance of essential services and Bushlight where each party agrees to work together, in a spirit of cooperation, to maintain and sustain the energy services. The CSA must clearly articulate the roles and responsibilities of each party as well as describing maintenance and repair and arrangements.[106]

The actual cost of installing a Bushlight renewable energy system depends on the configuration. All of the costings are reported on the website. One example from the Gunun Woonun case study (Mornington Island) reports a total cost of the energy system at $112,073. This included a PV array of 30 kW (40 × 75 W modules), 2,400 Ah @ 24VDC battery bank, 2.2 kW @ 40W and 24VDC inverter, 2 × 60 A @ 24VDC charge controllers and two service visits as well as connection of the generator, house wiring and lighting, energy management fittings and construction of the concrete slab. The Renewable Remote Power Generation Program (RRPGP) provided a rebate of $54,915 on the total cost. The renewable energy system subsequently saved $25,745 per year in diesel costs. It also saved 47.66 tonnes of greenhouse gases. With an approximate lifespan of 20 years for the system, even without the rebate, the cost of renewable energy would be approximately $5,500 a year as opposed to $25,000 for the previous diesel generator's annual cost.[107]

FIGURE 6.8 Uluru in Ulpanyali country in the Northern Territory, Australia. Photo by Meg Jerrard from Unsplash.

Grant funding is a great way to kick-start a community project; however, the funding and governance arrangements for the long-term sustainability of the technology must be established up-front, otherwise the systems will not be sustainably funded for maintenance, upgrades and expansion. In the Bushlight case studies, this issue was alleviated in three case studies. The first example was the Mabunji Aboriginal People, who set up a trust fund, which was used to maintain and upgrade the system as needed. The second example was the joint venture structure employed by CfAT under the Aboriginal and Torres Strait Islander corporation structure with a subsidiary engineering company, used to expand and fund new projects where grant funding and security of tenure was unavailable and community interests were varied and widespread across multiple interest groups. The third example was the Ulpanyali community that had continuous funding from income from the tourism at Uluru. This income went into a trust to fund ongoing infrastructure needs.

These are not the only structures that can be used. A hybrid funding system, using a grant up-front and any funding or governance structure that enables ongoing sustainable funds for maintenance, upgrades and expansion would suffice. Options will be explored further in Chapter 7.

A primary consideration is the workforce. Localised services are needed with local training and employment to negate the issue of a transitory workforce for maintenance.[108] This has been addressed by CfAT through joint ventures with

MyPathway, Universities and TAFEs. The Bushlight programme also gave ad hoc training to individuals as needed in some instances.

In summary, grant funded programmes should not be solely relied upon for economically sustainable essential services, due to the lack of reliability of a source of funds to maintain and upgrade infrastructure. While a grant can be useful to initiate a project, it should be supported by a governance system, such as a trust fund with income, for the infrastructure to be sustainable in the long term.

Where a sustainable governance model has been established, grants can be applied for through the Aboriginals Benefit Account,[109] The Indigenous Business Association or the Outback Power Program.[110] Aboriginal Land Councils can also provide grants for infrastructure for their communities. Infrastructure funding can come from a variety of avenues. For example, the Australian government's National partnership on remote Indigenous housing or the Northern Territory's initiative, Working Future. The Federal Department of Families, Housing, Community Services and Indigenous Affairs ARENA is also proactive in community grant funding.[111] Most State governments have funding available for community infrastructure.

It is important to consider the benefits of partnering with public bodies for grants as they can also assist with building new detached zero emissions housing for communities. There are now detached dwelling houses available that combine design, insulation, energy saving appliances and PV solar panels to create zero net energy housing. Often the detached dwellings produce more electricity than they consume.[112] For example, in the Kakadu case study, the energy infrastructure was partly funded by the Federal Department of Families, Housing, Community Services and Indigenous Affairs. In such instances, a comprehensive approach to the housing needs should be considered when designing the solar infrastructure.

Despite the perceived barriers to sustainable energy infrastructure investment in remote communities, the case studies above demonstrate numerous innovative solutions that have been used across the top end of Australia. When read in conjunction with the other case studies, a myriad of opportunities become apparent.

Pacific Island Indigenous and rural communities

Pacific Renewable Energy Project

1. Jurisdiction/region

This case study covers 12 countries in the Pacific Islands: Palau, Federated States of Micronesia, Marshall Islands, Kiribati, Samoa, Timor-Leste, Papua New Guinea, Solomon Islands, Vanuatu, Tuvalu, Fiji and Tonga.[113]

2. Background of community participants

FIGURE 6.9 A small solar installation in the Pacific Islands. Photo by Jeremy Bezanger from Unsplash.

Projects were completed in rural and indigenous villages and communities that identified issues with electrification. The majority of locations had populations below 200,000 people.[114] There are no local fossil fuel reserves in the Pacific, other than in Papua New Guinea (PNG). The locations are all remote and the cost of transporting diesel for fuel is high.[115] Many of the locations had access issues for receiving diesel transports and maintenance of diesel generators on a regular basis.[116]

3. Technology used

The most successful technology installed was small-scale solar PV, see, for example, Figure 6.9.

Some examples of technology used in Pacific Islands:

a. Kiribati uses small-scale solar and grid connected solar. This powers health centres, telephone sites and homes.
b. In Fiji, a village cooperative owns small-scale hydro that powers over 200 homes. Geothermal is also being developed. Solar has also been installed.
c. Papua New Guinea has installed hybrid renewable energy systems incorporating wind and solar.
d. Samoa has medium-scale hydro.
e. The Cook Islands uses biodiesel made from coconut oil.
f. Vanuatu uses coconut oil biodiesel for buses, taxis, cars, generators and hydroponics.
g. The Solomon Islands has installed hydropower.[117]

4. Funding model

The Pacific Renewable Energy Project started in 2000 in Australia. Austrade and France with the help of the World Bank started investing in renewable energy in the Pacific Islands. The funding was a combination of grants and loans.[118] The Asian Development Bank, part of the World Bank group, recently granted US$6 million in Tuvalu to help the island nation become 100% renewable.[119]

The International Finance Corporation, part of the World Bank group, is now trying to attract private sector development investment in renewable energy and the Pacific Islands.[120]

5. Legal/governance model

Many locations had limited regulations and governance frameworks.[121]

Grants and loans from foreign governments or the World Bank were the foundation of all projects.[122]

Public-private partnerships are now being sought.[123]

6. Benefits

The Asian development Bank is owned by 67 member countries, including Australia. One of its principal objectives is to reduce poverty in the Asia Pacific region. This includes funding for infrastructure development. The bank approves over $30 billion in infrastructure funding a year and employs over 3,000 people. This makes it a prime example of a lending facility for developing communities in the region.[124]

The bank has already invested in over 13 projects with over $400 million. There are further 16 projects in the planning stage with over $1.5 billion in investment.[125]

Austrade has identified increasing opportunities for private sector investment in this region. The main opportunities for investment are in independent power producers that are grid connected solar plants and can benefit from FiTs. An increased investment demand has been identified in renewable energy design and implementation, battery storage and grid integration, feasibility studies and integration with utilities. Technology suppliers are also needed as well as contractors. It has been identified that there are many similarities in the Pacific projects to projects in remote Australia.[126]

7. Costs

Many issues and barriers unravelled in the beginning of the Pacific Renewable Energy Project. Most of those issues have now been resolved. They have been listed below in summary:

a. Low political priority in the various countries;
b. Lack of local policy in relation to electricity or renewable energy;

c. Lack of energy infrastructure planning;
d. Management of village level politics and international private companies met with some difficulties;
e. Lack of building codes for energy efficiency;
f. Lack of maintenance procedures training;
g. High transaction costs;
h. Financial sustainability of loans from the World Bank as there were some issues with repayments;
i. Supervision of financial donations as it was reported that they were not put to efficient or organised use in some instances;
j. Grant funded projects created a low sense of ownership among locals;
k. Education and training in maintenance needed to be improved as well as retention of trained staff;
l. Wind turbines were environmentally difficult to approve due to the fragile nature of reef systems and cyclonic weather; and
m. Coconut oil as a biofuel requires a copra industry as well as indirect injection engines because of the high viscosity of the coconut oil.[127]

Solar PV systems were found to be the best systems requiring less maintenance than diesel generators and outperforming the other renewable energy infrastructure.[128]

FiTs and soft loans that were able to be paid off through electricity generation were found to be the most sustainable solution to the issues listed above.[129]

8. Learnings and recommendations

As a result of the Pacific Renewable Energy Project communities in the Pacific Islands now have reliable affordable electricity. Shops are able to open at night and streetlights are available in remote communities. Households can be productive in the evenings, for example, studying. Hygiene and health has improved through the use of electric appliances such as washing machines and fridges. Communication has improved through television, mobile phones and radios. Health issues associated with biomass burning such as lung disease and burns from open fires and kerosene lamps have been prevented.[130]

The Pacific Islands Renewable Energy Project has many similarities to remote Australia. The issues of access and cost of diesel supply have been resolved through renewable energy projects. The communities have benefited from renewable energy electrification in reliability, affordability and health.

Indigenous and rural Asian communities

Solar

Grameen Shakti, Grameen Bank

1. Jurisdiction/region

Bangladesh and India. In Figure 6.10 the solar installation powers the Bangladeshi farmer's irrigation system which saves on the cost of diesel fuel.

2. Background of community participants

This case study looks at agricultural communities and in particular the local women living in poverty.[131]

3. Technology used

Grameen Shakti is an offshoot of the Grameen Bank and focuses on renewable energy businesses, in particular solar panel systems. It was started in 1996. By 2007, the company was one of the largest suppliers of solar systems, having installed 100,000 in Bangladesh. By 2017, they had installed over 1.8 million systems.[132]

They also provide biogas technology that reduces stove pollution and converts natural waste such as animal manure into fuel for cooking.[133]

4. Funding model

Microcredit/microfinancing

The Grameen Bank lends out over US$2.5 billion a year to women in poverty to help them start a community business. Loans are made on the basis of trust, there

FIGURE 6.10 A farmer in Bangladesh stands in front of a rural solar installation. Photo by Mahfuzul Bhuiyan from iStock Photo, Getty Images.

is no security and no formal requirement to repay the loan, yet to date 98.6% of loans have been repaid. This is a financial innovation created by the founder of the Grameen bank, Mohammed Yunus, in 1976 to make capital available for women in Bangladesh.[134]

Yunus describes of a world of three zeroes,

> They are zero poverty: an end to income inequality; zero unemployment which is related to conceiving of people as job creators; and zero net carbon which is about preventing climate change but also about aligning the economic system to sustainability.[135]

Yunus talks of the impact of microcredit in enabling millions of people in India access to services, such as energy, which they would not have otherwise been able to afford. He analyses the shortcomings of a traditional banking system that will not easily lend to unsecured debtors.[136]

Lending to women in communities is a viable investment opportunity:

> When Africa's women farmers thrive, everyone benefits: the women themselves, the children in whom they invest, the communities that they feed, and the economies to which they contribute. With the right investments and policies, Africa's woman-run farms could produce a bumper crop of development.[137]

Yunus has also helped develop the "Social Success Note"[138] through the Yunus Social Business (YSB) and the Rockefeller Foundation.

> It is an innovative financial instrument that uses results based financing. In this system, a government agency or a charity organization underwrites loans from private investors to a not for profit organisation (NFP) that wants to start a project to pursue some specified social goal. If the program created by the meets agreed performance targets, the government will provide funds that make possible bond-like return on the loans. This strategy has been used successfully to attract funding for social programs from private investors like Goldman Sachs.[139]

Yunus says,

> I launched the bank without having any ambitious goals; I simply wanted to make life a little better for poor women in the villages of my home country. But over the past decades I have increasingly found myself engaged in redesigning the economic engine and trying out the new model in the real world.[140]

5. Legal/governance model

There are no legal documents involved in the loan arrangements. All loans are made on a trust basis. The system is also collateral free, aimed at women with no assets.[141]

> Grameen America survived the financial crisis. Apparently the integrity and hard work of women living in the villages of Bangladesh and the inner city of New York make a more reliable basis for lasting economic value than the clever constructions of financiers. A similar situation had occurred in 1997. The macroeconomies in a number of Asian countries declined steeply when a bubble of speculative lending burst, but the microfinance organizations in those countries continued to thrive. It seems that during an economic crisis, microfinance organizations can be an island of stability while "mainstream" financial institutions totter.
>
> *Yunus*[142]

The positive incentive to repay the loan is continued access to credit and mutual support to ensure loan repayments: "Grameen Bank has never used lawyers or courts to collect any of its loans .... Interest rates for loans and savings are clearly available for all to see on Grameen's website."[143]

The loans are given for activities that produce an income, housing and education, not for consumption. "The basic interest rate for most business loans is 20 percent on a declining basis, with no compounding; this is below the government-fixed microfinance interest rate of 27 percent."[144] This is still a high interest rate compared to a secured loan.

Grameen offers interest free loans without time limits to approximately 100,000 struggling members to encourage those members to cease begging and to become regular savers and borrowers. "A growing number of these borrowers have left begging behind completely and become door-to-door salespeople or adopted other income-generating activities."[145]

To promote accountability and openness, the bank is 75% owned by the borrowers (members) and nine of its twelve directors are female borrowers who are elected by their fellow bank borrowers with the bank being profitable and self-sufficient, generating enough money to remain solvent and independent through its simple system of lending, loan repayments, and member savings.[146]

In contrast with the traditional banking system, where little direct communication exists between the borrower and the lender, "Grameen's bankers and borrowers meet and look each other in the eye each and every week during the centre meetings that are held in eighty thousand villages all over Bangladesh."

The Bank does not think that complex contracts are suitable for a healthy relationship with ordinary individuals or bankers who find them impossible to understand. Yunus,

> I'm not proposing that we should try to radically simplify the legal and financial systems of the developed nations, making them purely trust-based like Grameen Bank. I am saying that the legal and financial challenge of creating a

> whole new sector of the economy based on selflessness, sharing, and the quest for social benefit—and held together largely by mutual trust rather than formal sanctions—may not be as complex or daunting as you might assume.[147]

Yunus believes that lawyers can help by using this trust-based model in other sectors of society to help build "human, family, and social capital by helping the poor—especially poor women—to help each other in a voluntary and business like fashion that builds respect, self-esteem, and community."[148]

He states that the Reserve Bank and government can also assist. The Reserve Bank of India is now issuing limited banking licenses to successful NFP microfinance institutions allowing them to become microfinance banks.[149]

> When you are building an organization whose mission is not to enrich any individual but rather to help make the world a better place for those in need, most people are happy to support it in the same spirit of altruism. Competition among market participants seeking to outwit each other becomes unnecessary. Elaborate safeguards to prevent exploitation are less important than they are in the world of profit-maximizing business. As long as a clear separation is made between the realm of social business and the realm of traditional profit-maximizing business, both realms will be able to flourish. And as more and more people become familiar with the concept of selfless business, participate in creating social businesses, and enjoy the benefits they create, an understanding of "dog-help-dog" economics will spread. This will make it easier for people to work together in a spirit of mutual trust without the need for elaborate contracts to control their interactions.[150]
>
> *Yunus*

Accessible financial services are an overlooked form of social infrastructure that is vital for poor women, according to Yunus. "Financial services like credit, savings, insurance, investment funds, and pension funds create economic opportunity for people and ensure growth at all levels, which is why it's vitally important for government to ensure that such services are available to everybody." However, Grameen Bank demonstrates that this can be done by entrepreneurs, community members and banks with individuals taking social responsibility for their own communities.

6. Costs

Yunus explains the risk to entrepreneurs with investments in jurisdictions lacking strong democratic principles. When his bank became profitable, it was appropriated by the Indian government:

> Grameen Bank is self-sustaining, runs with its own resources, has a high rate of loan recovery, and is owned mostly by poor female borrowers. It promotes savings; provides insurance and pension fund services; facilitates

> entrepreneurship; and gives power, freedom, and dignity to millions of illiterate rural women. Grameen Bank's forty-year history of nonstop success helps to explain why the bank won the Nobel Peace Prize back in 2006. Given this track record, it is surprising that the governments and central banks of the world have largely ignored their responsibility to ensure that the poor have access to financial services. I am disappointed that global women's organizations have not adopted guaranteeing such services as a key item in their agendas for empowering women. Even more shocking is the way Grameen Bank is under attack by the government of Bangladesh. The governing law of Grameen Bank has been amended to convert Grameen Bank into a government-run bank, taking control away from the borrower-owners. The government has not even allowed the bank to appoint its own CEO six years after I was ousted from this position in March 2011. The things that are happening to Grameen Bank represent a big step backward for the world. Given the history of government-run banks in Bangladesh, one can easily conclude that Grameen Bank is now on a path to disaster. It is heartbreaking to see a history-making, Nobel Prize– winning institution that gave birth to the concept and practice of banking for the poor and inspired the whole world to find a new direction in banking being pushed to take a sharp reverse turn because of these drastic changes in its governing law.[151]

Instances such as this drastically disincentivise entrepreneurs and investors from entering into a market. Yunus now works predominantly with poor women in the US and Europe where democratic governance of the rule of law creates a trustworthy environment for the investor.

The recent trouble that the microfinance sector has undergone in India[152] has occurred after, "The governing law of Grameen Bank has been amended to convert Grameen Bank into a government-run bank, taking control away from the borrower-owners."[153] As quoted above, Yunus foresaw many of the issues that have now occurred in the microfinance sector in India, which could have perhaps been avoided if the Bank had not been compulsorily acquired by the Indian government. This is an example of the importance of democratic free-market principles being upheld for a system to be sustainable in the long term.

There has been one significant anthropological study of the effects of microfinancing on women in agricultural communities conducted by Aminur Rahman.[154] Some of the issues identified in the study that need to be rectified before it is applied in Australian remote communities include:

a. The high interest rates and unscrupulous practices of some local bankers have resulted in some loans being circular, that is, while it is against the banks policy, some local bankers have allowed borrowers to keep borrowing more money to pay previous debts.[155]
b. The volatile rural markets and lack of insurance make the earnings unsustainable.[156]

c. The positional vulnerability in the family unit suggests that women are sometimes coerced into borrowing money on behalf of male relatives but are left with the sole responsibility of repaying the loan. Rural women are vulnerable to the patriarchal ideology in their communities.[157]

d. Joint liability has replaced collateral and has caused some issues:
   - The regular requirement to attend meetings is difficult for rural women who have farm responsibilities.
   - The bank can transfer default risk from the institution to the borrowers because they are jointly liable.
   - The women who are in charge of local banks/loan centres come from different lineages which may be factional in the community.
   - The loan centres become de facto power dominators in the community.
   - Poor women are rarely able to fulfil the obligation of joint liability to pay defaulter's loans and maintain group solidarity.[158]

e. The factional group formation and hierarchies within the rural culture in Bangladesh have created a class system within the loans centres.[159]

7. Benefits

As a result of the appropriation of his Bank by the Indian Reserve Bank, Yunus launched a global foundation, Grameen Crédit Agricole (GCA) Microfinance Foundation, a collaboration between Grameen Bank and Crédit Agricole. This is a trust that funds microfinance organisations on a global scale. "Crédit Agricole endowed the new foundation with 50 million euros."[160] The foundation is now able to help even more women across the globe.

Competition is sometimes a disincentive to investors who might prefer a monopoly of the market before investing. However, since Grameen Shatki was established, more than 30 companies have entered the market in Bangladesh. Yunus welcomed the competition, reporting that an extra 1.5 million homes were supplied with solar power by his competitors, for example, Green Village Ventures, which provides rural households in Uttar Pradesh, one of the poorest states in India, with access to solar power.[161]

Despite the risks identified above, affordable renewable electricity has been supplied to millions of poverty stricken homes. The financial industry, as with every industry will have risks. In most western countries, we have a highly regulated industry so those risks will be mitigated if the recommendations below are followed.

8. Learnings and recommendations

The following risks identified in Bangladesh by anthropologist Aminur Rahman can be mitigated as follows:

a. Ensure lenders have a financial licence so that they are subject to banking regulations.

b. The lenders should require contracts, albeit these could be simplified, to ensure consumer rights, that are already regulated, are known to the parties and upheld.
c. Supervision by the bank to ensure the loan is being used for the benefit of the woman to whom it is lent and not being passed to a male in the family for use on consumables.
d. Insurance strategies should be considered.
e. Women should be prioritised in the development of proven energy infrastructure such as solar systems so that the loans to women have a higher chance of going towards sustainable businesses.
f. Joint liability should be disallowed.
g. Circular loans should be disallowed.

To mitigate the risk of appropriation by a current or future government, a Federal regulated asset base could be instituted with a guarantee against appropriation of the investments. This would ensure continuity.[162] For example, if an energy company builds a solar facility, it agrees a reasonable cost of construction with the regulator. The company makes the investment, and finances it through a mix of borrowing and income from customers. The debt is entered into the register showing a return on investment (ROI) figure against the income from customers. In return for registration, the assets are treated as in perpetuity, because they are services that will always be needed.[163] They are not subject to depreciation but a capital maintenance charge. "This way the company does not face the risk of pure marginal cost pricing and there is no decline in functionality. In other words, they are sustained."[164] This maintains the aggregate of the asset intact.

Yunus, the founder of Grameen Bank says,

> Most villages in Bangladesh are not served by the national energy grid. Those that are connected find the energy supply often interrupted by outages. And, of course, traditional sources of electrical power like gas or coal-fired plants contribute significantly to climate change, whose terrible impact on Bangladesh I've already mentioned. For all these reasons, bringing clean, affordable, reliable energy to the homes of some 12 million Bangladeshis is a gigantic step forward. It provides schoolkids with electric lights by which they can do their homework. It enables shopkeepers, community centres, doctors' offices, and mosques to extend their hours into the evening, enriching countless lives and expanding economic opportunities. It helps farmers irrigate their lands and use labour-saving tools; it makes power-driven sewing machines available to rural female entrepreneurs. And it helps millions of Bangladeshis use the Internet to get access to the same sources of information and knowledge that people around the world rely on. Just as the rural electrification program launched by the New Deal in the 1930s helped bring poverty-stricken areas of the American South into the twentieth-century economy, so the spread of solar energy is helping to integrate the villages of Bangladesh into the world of the twenty-first century.[165]

Indigenous and rural Australian would benefit from an innovative financial system such as microcredit for the same reasons. As discussed above, Indigenous Australians in remote areas are asking for reliable, affordable electricity so that they can refrigerate fresh food and medication, children can do homework and street lights can be lit for safety.

Biogas

1. Jurisdiction/region

Rural and agricultural communities, international comparison. In Figure 6.11 cattle farming and biogas production are complementary income streams for the farmers.

2. Background of community participants

Australia is in the fledgling stage of developing biogas plants.[166] During 2016–2017, 0.5% of the national electricity was produced by biogas (1,200 GW). There are approximately 242 plants, half of those are connected to landfill sites.[167] It has been suggested by the Clean Energy Council that this could be expanded to include

FIGURE 6.11 An example of a mid-sized biogas installation on a cattle farm processing cow dung. Photo by CreativeNature_nl from iStock Photo, Getty Images.

plants connected to agricultural waste, forest residue, paper mill waste and urban waste.[168]

India, Indonesia and China have been developing rural biogas digesters for agricultural communities since the 1960s, aiming firstly, to reduce energy poverty for rural communities and secondly, reduce the pollution caused by the burning of biomass. More recently, Germany has started manufacturing modern biogas plants for large scale energy production.[169] Across Europe, the sector has grown, more than tripling to more than 17,400 biogas plants. Germany is the leading biogas country in Europe producing 100TWh of biogas energy. The United Kingdom comes in second at 23 TWh of biogas energy.[170]

3. Technology used

Biomass burning is the traditional way of cooking. In many rural communities, traditional kindling or modern wood chips are burnt for heating or cooking. The smoke from the wood burning causes lung disease for the women and children who predominantly do the cooking in those communities. Additionally, when the surrounding areas are deforested, soil erosion occurs, destroying the agricultural base of the communities. Cooking smoke is the highest health risk factor above smoking and high blood pressure in India and South-East Asia. In those regions combined, air pollution from conventional wood stoves is responsible for 4 million deaths each year. The cost to the national healthcare system in India (not reflected in the price of wood or energy) is between $212 billion and 1.1 trillion. Women and children are most vulnerable to these hazards as well as the hazards of spending up to four hours a day collecting fuel. In addition to the pollution, there are many reported cases of women and children sustaining injuries while collecting fuel.[171]

Biogas plants have come to replace traditional biomass wood burning. Biogas is collected in a small-scale plant. Inside the plant, bacteria anaerobically digest the biomass and convert it into biogas. The gas is then burnt for cooking. There are two primary types of biogas plants:

a. Covered effluent ponds for liquid waste where the biogas is piped for processing.
b. Engineered digesters in fermentation tanks. The process is controlled with heating and cooling and adding bacteria.[172]

Feedstock for the biogas plant ranges from livestock or human effluent, to organic landfill and wastewater or decaying vegetation or food waste.[173]

The burning of biogas is not the best form of renewable energy as it still contributes pollution. Biogas typically contains 55–20% methane, 20–20% carbon dioxide and trace gases such as toxic hydrogen sulphide and nitrous oxide. Methane has 21 times the power of carbon dioxide as a greenhouse gas. Capturing and burning it can theoretically reduce climate change, as opposed to letting it be released into the atmosphere.[174]

As the burning of gas still creates pollution and warming, this technology is best used where solar and wind are not viable. For example, in national parks, where tree coverage prohibits solar capture and forest residue is collected for disposal, a biogas plant could be used instead of solar panels. Also, industries where large amounts of methane are emitted, such as the agricultural industry, could benefit from additional revenue for the disposal and processing of waste products.

The small scale two-three cubic metre plants are used for home cooking and heating. Larger industrial plants, such as those manufactured in Germany can be used by farms, businesses or even neighbourhoods.[175]

4. Funding model

The majority of small plants supplied to rural agricultural communities in India and China are funded through microcredit, low interest loans and donations/grants.[176]

Case studies in Australia have suggested that medium size biogas plants, as used on agricultural holdings, will not attract investment where:

a. Biomass is not available onsite (as in a landfill site or agricultural site) because the cost of purchasing or growing biomass for fuel is too high[177]; and
b. Where other sources of revenue (such as forest waste treatment, sale of energy, sale of processed digestate) do not exist on site.[178]

However, where those additional forms of revenue can be identified on the subject site, the project can achieve the required CAPEX (capital expenditure) and OPEX (operating expenditure) to attract investment.[179]

5. Legal/governance model

The effectiveness of the biogas plant comes down to governance. For instance, the company REAP (Renewable Energy and Rural Electricity Access), in Mongolia, funded more than 60 call centres to assist with warranty claims and maintenance. By comparison, the Village Energy Security Project (VESP) in India had approximately 50% of its installed plants break down within 2 years due to lack of maintenance, confusion over who was responsible for maintenance and insufficient warranties.[180]

Grameen Shakti in Bangladesh has a more effective programme than VESP; increasing employment, education and training in the communities where biogas plants are installed. It provides subsidies for education and employment programmes, as does the REDP (Renewable energy development programme) in China.[181]

Grameen Shakti, China's REDP, Mongolia's REAP, Nepal's REDP and Sri Lanka's biogas project were also successful in part because the companies instituted fines for poor performance or violation of the companies standards of installation and maintenance.[182]

One of the most successful biogas projects was developed by the Shaanxi Mothers Association in Northern China. They installed 1,500 biogas plants on farms across the Shaanxi province. Those farms are self-sufficient in their energy use. The project started because the province was suffering from deforestation as the wood was being used for cooking. The deforestation led to soil erosion and desertification, dust storms and decreased water in the river. There was significant air pollution and pollution from animal waste. The cost of food and energy was increasing. The local government started restricting the chopping down of trees and paying farmers to replant trees. The Shaanxi Mothers Association decided that an alternative energy source, biogas, would solve the environmental issues.[183]

Most households owned pigs and those pigs produced methane gas and refuse that polluted the nearby rivers. Biogas systems used animal, agricultural and human refuse to generate the biogas for energy use. The by-product from the biogas system was fertiliser that was then used by the farmers. The Shaanxi Mothers provided extensive communication and training to the farmers. The education programmes included environment and health, the food chain, renewable energy, waste collection, pigsty and toilet retrofit, pipe laying to transmit the gas, equipment operation, maintenance and repair as well as creating fertilisers and organic farming.[184]

They also promoted solar hot water systems and retro upgrades to toilets that integrated into the biogas system. The families were taught how to collect rainwater and do woodwork. The workshops included women and their children to encourage women's self-confidence and mother-child bonding. The villages were involved in the planning and construction of the projects. The systems last for 15 years if maintained correctly. They have a payback period of one year. The local government and the Shaanxi Mothers subsidised two thirds of the cost. The farms also save on wood fuel and fertiliser. Food production also increases on the farms to the good quality residue fertiliser. The cost of each unit is about $400–$500.[185]

Each biogas system is about 8 cubic metres and produces around 400 cubic metres of gas daily. This is the equivalent to 2,000 square metres of firewood. It also negates the need to transport coal to the community.[186]

In 2005, the Hong Kong charity called Friends of the Earth started to assist and created the Sunflower Project, aiming to subsidise biogas for every rural households in China. Following this, donations have come from, The Global Women's Cooperation, Global Biogas Capacity Building Project, Global Fund for Women (US), Badi Foundation for Village Capacity Building (Macao), The World Bank, Wufang Xiaowei, The German Consulate and local government agencies. The majority of those funds were spent on retrofitting pigsties, toilets and kitchens ($133,970). Only $16,560 was spent on administration. $32,940 was spent on training.[187]

For a household to receive a biogas subsidy, they must demonstrate the number of people it will benefit, the number of animals that will be kept, the conditions of the animals' space and they must be literate. The ownership of a biogas system or solar hot water system has become a status symbol in the region.[188]

6. Benefits

The key drivers for existing biogas plants have been:

a. Reducing odours from abattoirs or rendering plants; and
b. Generating electricity for landfill sites, and waste water treatment plants.[189]

Investing in anaerobic digesters can produce fertiliser, increase water efficiency or derive revenues from renewable electricity production.[190] Energy produced can be sold to the grid if negotiated with network providers.[191]

It has been suggested that biogas plants could potentially divert 2,051,200 Mg tonnes per year of landfill food waste into energy at approximately 1,915 GW per year assuming 50% of Australia's food waste is used.[192]

The benefits to rural and remote communities where solar and wind are not viable have been noted in particular.

The Shaanxi Mother's Association reports that the health of women and their children has significantly improved because they no longer have to breathe in wood smoke which causes respiratory and eye disease. Carbon monoxide emissions were reduced by 80%, CO_2 by 60%, SO_2 by 80% and dust and fumes by 90%. Sanitation of the animals and the human toilets also improved and waterborne diseases reduced. The density of flies halved. The organic fertiliser by-product improved crop growth by 5–10%. Over a million people have now been involved in the project. Deforestation has halted in the region and local income has improved as well as the overall standard of living. The status of women has also been improved and women have become more self-reliant.[193]

7. Costs

There are risks to the use of small biogas plants in rural areas as demonstrated by case studies in India, China and Indonesia:

a. Communities were not trained in maintenance in some parts of India and Indonesia and biogas plants leaked both slurry and gas, causing environmental damage.
b. Warranties were not upheld by some Indian and Indonesian installation companies.
c. In some Indian communities, the residue from the biogas plants was being used as fertilizer without processing causing heavy metals and pathogenic pesticide residues to enter drinking and food sources.[194]

The difficulties listed above are unlikely to occur in western countries. The industry is highly regulated at the planning stage.

8. Learnings and recommendations

Small-scale biogas plants used to fuel household stoves have been funded across India and China through low interest loans, grants and microcredit. This technology could also be used by households in regional Indigenous communities where solar and wind are not viable options and where wood stoves are causing health and environmental hazards to the household. The CAPEX of small biogas plants is low compared to solar and wind technology.[195] So long as sufficient installation warranties and maintenance training is in place, the OPEX should be kept to a minimum.

Mid-sized plants for agricultural communities will attract investment where additional sources of revenue are available to offset the CAPEX and OPEX. For instance, waste treatment, sale of energy and sale of processed digestate as fertiliser.

Overall, biogas is useful as a renewable energy source where large quantities of free biomass are available for processing on the subject site, such as a landfill or human or animal effluent. It is also useful as an alternative renewable energy source for national parks where tree coverage prevents solar capture, or any subject site where trees are to be preserved and there is too much shade to collect sufficient solar energy.

High quality maintenance is necessary to prevent ecological harm from leakage of slurry or gas. Planning legislation should be in place to ensure plants are built to high standards. Additionally, consumer law should protect communities where installation warranties are provided. Financial, environmental and social sustainability of biogas is based on free excess biomass in the community (as opposed to extra biomass being grown for the purpose of biogas plants).

Australian rural community solar

When looking at remote communities, all aspects of the community need to be addressed. When it comes to land and property rights, there are many layers of property rights that exist. The communities will be made up of not only indigenous people but also, pastoralists, mining entities, State, federal and local government interests, telecommunications interests, as well as other property owners. For instance, in rural communities, the National Federal Farmers Association (NFF) has been particularly vocal in relation to the supply of electricity and other infrastructure to remote communities, stating for example,

> Reliability and affordability are key for farmers – wholesale price spikes and outages can destroy annual returns, in some cases, in the space of a few hours. Electricity use is variable across agriculture depending on the intensification of operations, location and business structure. For some farmers demand is flexible while for others demand is driven by factors such as weather and access to water. Efficient investment in, combined with efficient operation

> and use of, electricity services is crucial for farmers, other consumers and the wider economy.[196]

The NFF reports that Australia's farming industry contributes $60 billion worth of food and fibre annually. The NFF is calling for a reliable and affordable energy supply and energy security for the agricultural industry.[197]

Irrigation is a particularly costly exercise when energy prices increase. For example, Central Irrigation Trust's pumping station at Loxton spent A$880,000 on electricity in 2010, and in 2017, they spent A$1.8 million. As Steve Whan, National Irrigators Council Chief Executive, said

> That's a 107 per cent increase. They can't go to the big supermarket chains and say 'Hey, our electricity costs have gone up, we'd like you to pay a bit more for your fruit and veg'. They're effectively price takers, so it's another thing that puts the economics of being in business on a farm at risk.[198]

Some farmers have switched from grid connected electricity such as Graham Clapham, who farms with his family on southern Queensland's Darling Downs, "Closely following the cost of the energy (as a problem) is the reliability of the network. The network in recent years has become quite prone to outages," Mr Clapham said.[199]

Other issues with unreliability and outages is that they can ruin watering schedules if pumps switch off without warning and do not restart themselves when the power comes back on and fruit growers risk losing their newly picked crops if they lose power in the packing shed and cool rooms over harvest.[200]

Jenna McGregor describes power reliability as "awful" on the property where she and her husband farm at Tenindewa, about 470 km north of Perth, "We had a shocking summer. Our power went out copious amounts of times," she said.

> On the majority of the hot days that came through, our power went out. Not just for an hour or two, sometimes 50 hours, sometimes 24. We lose our water because our pumps no longer function, we lose phone signal because our phone boosters no longer work. For people that don't have generators it can be very uncomfortable to live through.

Ms McGregor further added

> without power, the heat was uncomfortable but it also cost money. We are receiving a sub-standard service. We are paying a lot of money for power. We are paying the same amount that people in towns and cities are paying and we're not receiving the same service.[201]

Many farmers are installing renewable energy because they are frustrated at the unreliable and unaffordable grid connected prices. For example, Mary Nenke runs her yabby farm and eco-cottage business on solar and has been off the grid for more than a decade. The property was a sheep and grain farm but diversified into tourism and yabbies in the 1990s.[202] It cost approximately $250,000 to go off-grid with the farming and eco-cottage. They received a $30,000 State Government rebate. They also saved on Western Power connection charges of $70,000.[203]

Another example is the Dobra family's packing shed at their loose leaf lettuce farm. They used to spend $14,000 a month on electricity during the peak summer; however, this was cut to $7,000 immediately after the panels were installed. The monthly repayments on the solar panels are $2,500. This has taken the farm into a positive cash flow.[204]

Large organisations are now seeing solar farm projects as an economically justifiable solution. The Queensland University of Technology generates electricity in the Brisbane city centre through on-site solar PV arrays. For instance, in the week of from July 27 to August 2, 2020 15,211 kWh was generated on-site and 14.15 tonnes of carbon dioxide were saved in greenhouse gas emissions.[205]

Other entities are using regional solar farms to power urban infrastructure. The University of Queensland built a solar farm asset in the farming region of Warwick in 2020 that generates all of their urban electricity using their own regional renewable asset.[206]

Providing energy security and reliability for agricultural areas is being recognised as an economic opportunity across all sectors and with economic opportunity comes employment opportunity.

The case studies in this section are installed in rural Australia. The first three case studies are solar PV projects that have each used very different governance and funding mechanisms and are analysed in turn below. The fourth case study for rural Australia is a micro hydro project that showed a much higher level of long-term sustainability than the other case studies; however, the costs associated were also higher. The next two case studies compare successful wind farm projects in Denmark and outback Australia by looking at the differing funding and legal structures. The final two case studies are large on grid solar farms in agricultural Australian towns that demonstrate good profits exporting renewable energy to urban infrastructure such as the Sydney Opera House and Coles supermarkets. They also demonstrate the ability for solar to integrate with agriculture, in this case, a cattle farm. Figure 6.12 shows a similar outcome on a sheep farm.

Private and public funding: Clearsky Solar Investments (CSI)

1. Jurisdiction/region

Clearsky Solar Investments (CSI) seeks investment Australia wide and has delivered the following projects in regional NSW communities.[207]

a. Boggabri NSW Solar PV 15 kW
b. Mudgee NSW Solar PV 50 kW

FIGURE 6.12 A herd of sheep grazing around solar installation in a rural landscape.
Photo by Vincent Delsuc from Pexels.

c. Walgett NSW Solar PV 25 kW
d. Mildura NSW Solar PV 75 kW
e. Broken Hill NSW Solar PV 100 kW[208]

2. Background of community participants

CSI is a partnership between community members and private investors supplemented by a state government grant.[209]

3. Technology used
   a. 15–200 kW solar systems.[210]

4. Funding model
   a. Hybrid:
      i. Community investment and State grant.[211]
      ii. Community investors, less than 20 can invest in an NFP organisation. Any returns on investment go back to the investors.[212]

5. Legal/governance model
   a. Trust structure and NFP association.
   b. A loan between the trust and a commercial partner, Smart Commercial Solar.[213]

The commercial investor (CSI) loan the funds to the installation company (Smart Commercial Solar) in return for a specified $ amount per kWh generated. Smart Commercial Solar enters into a power purchasing agreement (PPA) with a customer and then builds the installation. The installation company retains ownership in the solar system. The customer obtains affordable clean energy. The investor obtains a guaranteed ROI.[214]

An information memorandum or managed investment scheme fundraising disclosure is required. A financial services licence or exemption applies.[215]

Cost-benefit analysis

6. Benefits

This model is successful where community involvement is not essential and in particular where members investing in the projects do not want to be involved in the delivery of the project. The ROI is the key factor in attracting investors with this model. A proven track record of successful projects Australia wide attracts investment in subsequent projects.

7. Costs

Community engagement is constrained by the governance structure. That is, community members are engaged usually through the process of investing or becoming a customer. Projects cannot advertise for investors so the local community and their networks must invest in the local projects. The legal requirements under the *Corporations Act 2001* mean that there is a limited time in which funds can be raised. There must be sufficient investment found through during the limited timeframe. In this particular project, some of the funds required were granted by the NSW government to make up the shortfall. ClearSky also has a benefactor that contributes to the shortfall of many projects.[216]

This method requires extensive legal and accounting work, therefore pro bono professionals are required, or a method to fund the transaction costs must be accounted for in the planning stage.[217]

8. Learnings and recommendations

Transaction costs for community projects are often covered by pro bono work contributed by large firms.[218] However, this cannot be relied upon and transaction costs need to be considered in any project.

CSI engages with communities Australia wide. They engage with small-scale investors in their local communities who have access to a host site, not necessarily their own.[219] The small-scale nature of the projects has some limitation when attracting investments, such as restrictions on advertising and time limits for raising money. By contrast, the small-scale means that many of the onerous legal requirements of larger projects are foregone.[220]

This model would suit a community that preferred a solar installation company to retain ownership and along with it the maintenance and upkeep obligations of the solar PV. All obligations would need to be specified by various contracts and access to legal and accounting advice is essential. The community would ideally have their own investors that were interested, have access to grant money or have formed a relationship with ClearSky and their investors.

Revolving fund: CORENA

1. Jurisdiction/region

CORENA is an Australia-wide revolving fund with projects in New South Wales, Queensland, South Australia, Victoria, Tasmania and Western Australia.

2. Background of community participants

CORENA is a group of community minded benefactors who wish to contribute to the common good and reinvest the profits of renewable energy projects into further projects. Investors do not receive an ROI. Returns go towards the next project. The majority of the loans go to non-profit organisations.[221]

3. Technology used
   a. 5–20 kW solar systems.

4. Funding model

The initial funding is based on donations and grants. This money is then placed in a revolving fund. Subsequent donations, plus the loan repayments from the initial solar PV projects go into the revolving fund. All profits go towards reducing power bills to the host sites and to fund other solar PV installations.[222]

Fundraising activities are required to raise the donations. Charities can sometimes use donations to secure loans, if they are an established charity and have regular donations.[223] However, this is not an easy feat. The revolving fund, however, allows a charity to use donations to secure a loan.

CORENA provides interest free loans to non-profit community organisations to pay for installing clean energy systems (solar panels, batteries, solar hot water, and/or efficiency measures). The loans are repaid into the revolving fund and used to fund more projects in other communities.[224]

Due to the revolving fund structure, the donated amounts have approximately five times the impact of a normal donation. The impact of the donation is what attracts donors to these revolving funds. For example, $100 lent in the first

CORENA project in November 2013 has been re-lent to subsequent projects totalling $225 of funding revolving exponentially.[225]

Australia-wide payback periods were able to be compared across a wide range of projects. It was found that the payback period varies depending on different solar prices, electricity prices, solar radiation availability and FiTs.[226] The payback period is shorter for a system that is mostly used in the daytime.[227] It was recommended that a five-year payback period be set for Adelaide, Brisbane, Canberra, Perth and Sydney. A seven-year payback be set for Darwin and Melbourne. An eight-year payback be set for Hobart.[228] It was also found that a 5 kW system is the most cost-effective. Anything smaller was not recommended. If a householder has low electricity use, the savings on their power bills might not be enough to cover the repayments especially if the revolving fund charges interest.[229] CORENA does not charge interest on the loans but other funds might do so.

5. Legal/governance model

An NFP association is the vehicle used to enter into loan arrangements. The loan agreements govern the arrangements. A financial licence or exemption could be required.[230]

Cost-benefit analysis

6. Benefits

The revolving fund removes barriers to finance. ROI is not required to secure a loan. Community need is used to assess the loan application. Projects can be implemented by any incorporated association.[231]

According to CORENA, if a solar project has a payback period of five years (as do many of the CORENA projects), an organisation that starts a new project each year with a revolving fund can afford to fund one new project per year forever and after five years each project will be funded out of the revolving fund without need for any further donations.[232]

Within five years, CORENA raised $177,000 in donations. Twenty-three solar projects received interest-free loans. $133,000 in loan repayments went into the revolving fund. $310,000 worth of solar projects were built (202 kW). This saved approximately 446 tonnes of carbon emissions. This would be the equivalent to approximately a 75% ROI.[233]

The other benefits discovered by CORENA were:

a. Growing local economies.
b. Divesting from coal and gas.
c. Divesting from coal seam gas. This is an important consideration for farming communities who are worried about the pollution to the water table caused by fracking.

d. Communities who want to tackle climate change felt empowered.
e. Although the donations can't be used as carbon offsets for tax purposes, they are often made as moral offsets. For example, a local donor will donate to the revolving fund when they take a flight.
f. Regional credit unions use revolving funds models for large scale solar.
g. The revolving funds spread the benefit around many community groups.
h. Some projects install more kilowatts than is needed so that once the loan is paid back, the extra electricity produces an income for the community group.
i. A micro grid can be created in the local area to export the excess electricity to someone who would otherwise be using diesel or coal fired electricity.[234]

7. Costs

The community organisation that borrows the money from CORENA is the owner of the infrastructure and is therefore responsible for ongoing maintenance, operation and upgrades. Depending on the organisation, this could either be a sustainable or unsustainable endeavour.[235]

There is no ROI for the donor and donations are made for the common good. This could be seen as a cost or a benefit.[236]

The community organisation would ideally own the subject site. Long-term leases are also possible with the permission of the landlord.[237]

The ability to repay the no interest loan is considered in the loan application so that donors can maximise the benefits of their contributions by seeing the repayments revolve through the fund.[238]

A revolving fund is more difficult to initiate and gets easier as time progresses. The first project requires all of the CAPEX and OPEX to be donated. Once the loan from the first project starts to be repaid, those funds can be used towards the second project and so on.

CORENA focuses on projects it calls "big win" and "quick win" projects to give the donations a better chance to revolve.[239]

8. Learnings and recommendations

This model is suitable for a charitable organisation. The legal governance is simpler than some other models. If an established revolving fund such as CORENA or COREM is used to borrow the funds, the legal work is simpler than if the organisation were to start their own NFP and revolving fund.[240]

The essential requirements for a viable revolving fund project are:

a. A series of project hosts with suitable roof space and regular daytime electricity use;

b. A mechanism for fund raising, and a structure for depositing donations, issuing loans, and receiving loan repayments;
c. Access to professional product and installation advice to ensure high project quality;
d. Accredited installers and quality warranties;
e. Staff or volunteers to promote project fundraising, manage and report donations and loan repayments, and assess and liaise with prospective project hosts; and
f. Transparent financial and project reporting and annual financial audits so that the public and potential supporters can see how donated money is used;
g. Annual financial audits.[241]

According to CORENA, regional renewable energy donors prefer projects where a "quick win" is shown. However, they also state that a larger investment-based project might attract investors with the right publicity and governance.[242]

Revolving funds have since been established by several other individual communities such as the Mullumbimby COREM project which was established by the community as an alternative to the coal seam gas industry in the region.[243]

The COREM project partnered with an energy retailer called ENOVA which has a higher than usual FiT for residential community organisations. The community pays back the loans to the revolving fund faster due to the savings in the electricity bills.[244]

The revolving fund model alleviates the issue of donor fatigue. When fundraising in a small community, COREM found that the first project, solar PV for the local Drill Hall required a huge fundraising effort, $65,000 was raised. Subsequent projects did not require the NFP to spruik for more donations in the community as the revolving fund model is economically self-sufficient. COREM has since funded seven projects. COREM uses an MOU rather than a loan agreement. The biggest challenge for COREM was the permission and planning approval from the body that owns the property or site. Tenure issues and shading issues also occurred.[245]

Local councils are also trialling revolving funds. Solar saver is a revolving fund used by several local councils.[246]

An Aboriginal and Torres Strait Islander corporation could borrow from CORENA or establish their own CORENA style fund specific for Indigenous communities.

The primary benefit of this model is that the corporation could use the model to build infrastructure across a number of communities and after the initial grants or donations, the fund could sustain itself in a financially Sustainable System without relying on further donations or grants.

Additionally, an Aboriginal and Torres Strait Islander corporation could set up a revolving fund by partnering with the Indigenous Business Association[247] or National Indigenous Australians Agency[248] if assistance with governance, initial grant money and technical expertise is required.

Private: Repower One-Five

1. Jurisdiction/region

Repower started with the Shoalhaven Heads Bowling and Recreation Club (NSW) and has since completed four other solar projects including two dairy farms. They focus on isolated regions with high electricity prices.[249]

2. Background of community participants

Repower is Australia's first known community investor owned solar power system, initially started by local investors in Shoalhaven Heads, NSW.[250]

3. Technology used

Repower started with a 99 kW solar power system on the roof of the Bowling Club in the first project.[251]

4. Funding model

The Bowling Club purchases electricity from Repower One through a ten year PPA. Repower forecasts its returns at fully franked IRR (internal rate of return) at 5.26% per annum. The minimum upfront investment was five shares at $5, 990.[252] The Bowling Club owns 20%. Community investors can be shareholders or lenders.[253] Community investors contributed $119,800. Additional funding was received from community donations and NSW government grants.[254] Additionally, the solar installer provided a community-financing offer for installation.[255]

5. Legal/governance model

Repower uses an NFP organisation and proprietary company (also known as a special purpose vehicle or SPV) structure. The company is made up of community investors. Under the *Corporations Act 2001 (Cth)*, they must have less than 20 investors per year per project with a project maximum of 50 investors per project.

The model has a seven to ten year term in which the SPV owns and operates the solar farm. A PPA is in place with the host site. At the end of the PPA, the host becomes the owner of the solar farm. There is an option to purchase the system early at a mid-term buy-out value. Repower Shoalhaven is the parent community organisation and is responsible for metering electricity and invoicing the host for consumption under the PPA.[256]

Cost-benefit analysis

6. Benefits

This model has been repeated in five successive projects, Repower One through to Five.[257]

Due to replication, the cost of the project decreases with each subsequent project.[258]

7. Costs

One of the criticisms of this model is that an SPV is not an entirely independent community entity, "While the SPV is entirely owned by community members who are shareholders, it is governed by the board of the parent community organisation through a special shareholding that gives them voting power but no dividend rights."[259]

Another criticism is that there is a limit on the number of investors due to the SPV structure. The financial capabilities of the investors restrict the expansion of the entity into other projects. Sometimes bridging loans are used to address this issue. In one project, a significant investor underwrote the project.[260]

The SPV and the community investors are liable for the risks taken. As the owner-operator of the solar farm, the SPV (community investors) carry the risk of something going wrong with the performance of the solar farm.

The host site's financial health is important as there is a risk of a host site customer experiencing financial difficulty and being unable to make their regular payments and/or buy out the solar farm at the end of the term.[261]

The project has relied upon sweat equity (volunteers) which, if assessed at $50 per hour, amounted to around $90,000 in value.[262] It was recognised that professionals such as lawyers, accountants and a qualified chairman were necessary to the success of the project. These professionals might need to be pro bono, volunteer investors or volunteer members of the project to make the project viable.[263] After completion, the project's compliance costs accounted for 13–25% of the ongoing administrative monetary costs.[264]

The costs of the following also need to be taken into account; fundraising (equity and debt), tax, other legal costs, compliance and auditing requirements, financial licences, insurance, operating costs, structure set-up and operation.[265]

8. Learnings and recommendations

The project needs to fit within the following parameters to be viable:

a. The value of electricity saved (retail price) is much higher than the price for selling electricity sold through the electricity grid or network (wholesale price or feed-in price). This model works because the host sites use a significant amount of electricity during sunlight hours every day of the year. The solar farm is able to purchase all of the electricity at the time it is produced, so it does not need to be sold back to the grid.[266]

b. A system that is less than 100 kW has less risk on investment, keeping in mind that much of the income comes from trading the small-scale renewable energy certificates.[267] It is important that there is certainty when selling the renewable energy certificates generated by the solar farm either up-front (STCs for renewable energy generation of less than 100 kW in size) or throughout the period of the project for projects (LGCs projects larger than 100 kW). This is because the renewable energy certificates account for approximately one-third of the project's up-front capital or ongoing income respectively.[268]
c. Available professional support (volunteer or paid) to ensure governance compliance.[269]
d. Due diligence in all aspects of the project is important. For example, the installer, the technology used and the solar contract conditions. The installer, for example, was subject to the following due diligence:
   i. Tier One (BNEF list).
   ii. Having a third party warranty insurance provider (PowerGuard).
   iii. Having multiple layers of security in the supply chain (i.e., imported by a large wholesaler separate to the installer and a manufacturer who has honoured warranty claims in the past).
   iv. The installer verifying that previously installed modules are performing at least to expectation.
   v. The warranty of all products was set to last at least the length of the contract term, including 10 years inverter warranty and 15 years frame and racking warranty.
   vi. Remote monitoring capability with automated email alerts.
   vii. Ongoing support and fast incident response time was important to Repower Shoalhaven when selecting an installer.
   viii. After each project was installed, an independent solar professional conducted an inspection to check for any issues. Some minor issues were found and rectified.[270]

Council-community partnership: Lismore community solar farm

1. Jurisdiction/region
   a. Council owned and operated sites in a rural, agricultural community, Lismore, NSW.[271]
2. Background of community participants
   a. Local community members and the Lismore City Council.
3. Technology used
   a. Less than 100 kW solar systems.

4. Funding model

The funding model consists of a council-community partnership with an unsecured loan between the local council and the private company. Community investors bought shares in the private company. Fully franked dividends were returned to the shareholders. Investor's capital was returned at the end of the project.

The investor's rate of return depends on the rate of interest on the loan and the repayment terms (for example, interest only or principal and interest).

5. Legal/governance model

The community investors are governed by the private company (SPV) rules and requirements under the Corporations Act Cth *2001*.

The local government must have the legality of their model approved by the Office of Local Government.

A loan agreement has regulatory requirements.

Cost-benefit analysis

6. Benefits

As all of the host sites are owned by the council, the responsibility for operation, performance, maintenance and upgrades of the solar systems falls with them.

7. Costs

The ROI is linked to the loan repayments by the council rather than the price of electricity under a PPA, as in other comparative models. This could be seen as an advantage or disadvantage depending on the price of electricity and the loan terms.

Clearance is required from the local government oversight authority.

Shareholder limits for private companies reduce the amount of investors, therefore investors need to invest a greater amount of capital.

The Lismore solar farms required a minimum investment of $9,000 to raise $180,000.

8. Learnings and recommendations

This model is suitable for a local council who wants co-funding and community involvement in renewable energy projects.

Host sites, such as council buildings, that consume electricity all of the time make the project viable because the value of the electricity saved at the retail price is much higher than the price for exported surplus electricity sold through the grid at the wholesale price.

Professionals with relevant experience are required to be directors of the company. Investors must be able to raise significant capital. This model requires time, professional assistance and financial backing.

The benefit to the host site is the use of renewable energy and the improvement to the public perception and community engagement. More affordable electricity is another benefit.

This model has been adapted for use by Starfish Initiatives and Farming the Sun and has been applied to other community organisations such as Climate Rescue of Wagga and Manilla Community Solar Co.

Rural wind energy

A comparison between Australian and Danish rural case studies

1. Jurisdiction/region

The following is a comparison of two wind energy projects, the first, Hepburn Wind is an Australian project built in the rural community, Daylesford, Victoria.[272]

The second is the Danish region, Samsø, which was the first island in Europe to achieve 100% of its electricity from renewable energy, predominantly wind including EVs powered by wind energy.[273]

FIGURE 6.13 White wind turbines on a hill surrounded by mist. Photo by Fabian Wiktor from Pexels.

2. Background of community participants

This Hepburn Wind project is owned by a cooperative consisting of 2,002 local members. The community's industry consists of farming, such as, cattle farms, olive groves, vineyards and permaculture. There is also a tourist industry. The wind turbines are situated on a cattle farm.[274]

Samsø is a small agricultural town in Denmark.

3. Technology used

In Hepburn, two RePower wind turbines totalling 4.1 MW, powering 2,000 houses, were installed by Hepburn Wind. A small renewable energy developer, Future Energy, supported Hepburn Wind throughout the project.[275]

By 2005, Samsø Energy Supply Company had become an exporter of wind energy. The region was creating more energy than it used, made up mostly of wind energy and backed-up by biomass heating (wood chips) and biogas, produced from landfill sites. Upgrades to insulation and energy efficiency also assisted. About 300 individual homes also invested in renewable energy systems for heating.[276]

4. Funding model

The capital to fund the Hepburn Wind project initially came from a combination of issuing shares in the cooperative, loans and grants: A $2 million grant provided by Sustainability Victoria and Regional Development Victoria, $9 million was raised from members of the cooperative through the sale of shares and $2 million in debt to a local bank. Now Hepburn Wind generates income from selling electricity to the grid and Renewable Energy Certificates.[277] In order to encourage community ownership, Hepburn Wind has a policy of maintaining that 50% of the cooperative's membership self-identify as local people, with a minimum share offering of 100 shares to locals (valued at $100) versus 1,000 shares for non-locals (valued at $1,000). They also made an informal commitment that no single shareholder can hold a controlling interest (more than 50% of shares).[278]

Hepburn wind guarantees at least $30,000 to the community fund regardless of surplus or deficit. Their contract with the energy retailer helps with this arrangement. They have funded 36 community initiatives.[279]

Eleven small turbines (1 MW) were erected on Samsø Island at the cost of 8.8 million euros. Shares in the company were sold to the community members who received dividends from the sale of the power. Forty private applications for wind turbines were submitted so not all projects were approved. Some offshore wind generators were also funded in later years by the Danish Energy Agency at the cost of 33.3 million euros.[280]

The Samsø project was predominantly funded by individual investors. In the early stages of the project, no funding or tax benefits were given. However, as time went on funding and grants were attracted from the Danish Energy Agency, Aarhus Regional Authority and Samsø municipality, the Samsø Business Forum, Samsø Farmer's Association, Samsø Energy Supply Company and Finland's Ministry of Energy.[281]

5. Legal/governance model

Hepburn Wind is governed by a cooperative with member investors.[282] Local control of the project was important to the community investors, surprisingly as it is a large risk for a small community to take on.[283]

A cooperative does not need to prioritise the profit of the shareholders and this was the reason Hepburn wind chose this structure.[284] However, the difficulty with this structure is that they cannot sell electricity to the members. This is due to complicated electricity market regulations. By contrast, a fishing or grocery cooperative means that an active membership is sufficient to enable a person to buy goods from the shop and gain a corresponding benefit through reduced cost of the goods.[285]

Hepburn Wind chose to become a certified "B-corp." B-corps are enterprises that "voluntarily hold themselves to a higher levels of accountability" across governance and environmental and social impact. They sign a contract in which they agree "to the extent permissible under Australian law, to consider the impact of its decisions not only on shareholders, but also on its employees, customers, suppliers, the community, and the environment." To become a B-corp, Hepburn Wind undertook an assessment process to evaluate against a range of criteria aimed at ensuring higher standards of ethical operations. They are required to report annually.[286]

The Samsø Energy Supply Company (a private company) was established in 1998 to coordinate the various wind energy projects and the financing of the projects. Due to its success, many public agencies were subsequently established to engage with investment in renewable energy investment on the island.[287]

Cost-benefit analysis

6. Benefits

The wind farm, Hepburn Wind, paid approximately $15,000 per annum to lease the farmland. To avoid conflict among neighbours who didn't receive the lease payments, 1,000 free shares were offered to any neighbour within 2.5 km of the project. This solidified community support for the project.[288]

Hepburn Wind also offer $200 a year towards cheaper electricity for neighbours and have made the neighbourhood a priority in the distribution of community grants that are funded by the project to encourage small businesses in the area.[289]

The grants are organised through the Hepburn Wind Energy Fund which was established by the cooperative to fund other renewable energy projects and community minded projects in the region.[290]

The Samsø project created a local renewable energy industry as well as an income from exported electricity. Infrastructure was built as a consequence, such as a renewable energy research laboratory and Energy Academy as well as several public agencies: the Samsø Energy Agency, Samsø Energy and Environment Office and Energy Service Denmark. Eco-tourism also boomed with an eco-museum being built to accommodate the demand from tourists.[291]

7. Costs

Initially, the Hepburn Wind project was put forward by a developer and rejected by the community. The community then formed their own cooperative to build the project. Without community support, it is difficult for an investor to develop renewable energy infrastructure, especially wind energy.[292]

The feasibility of a wind project can vary depending on the initial costs. There are large legal, engineering and accounting costs. The planning stages can take one to two years.[293]

Smaller wind turbines have been found to be more effective for communities in Denmark such as Samsø Island. Because wind turbines have a lifespan of 18–20 years before they need upgrades or replacement (known as the repowering phase), smaller wind turbines are more cost-effective for communities to repower at the end of the lifespan. Large turbines that have been built in other parts of Denmark are more likely to be decommissioned. This is because most towns have become energy exporters, due to the abundance of free wind energy, the price for electricity is so low that it is unviable to repower a large 20-year-old turbine, on the other hand, the small turbines built on Samsø Island are more affordable to upgrade or replace and are preferable in the repowering phase.[294]

8. Learnings and recommendations

As with other renewable energy projects, the choice of property is key to a successful wind project. For example, the property should be checked for tenure, planning impediments or endangered species, access to suitable grid connection options, wind speed, construction access, and any visual and noise impacts.[295]

Sweat equity volunteers were necessary for these projects; however, it was seen as a positive aspect of the projects:\

> With a group of volunteers … they have a great deal of ownership over their opinions or beliefs … with the community model, everyone is brought in and everyone will see it to the end because everyone cares. That's the real strength.[296]

The projects also prioritised local suppliers and contractors wherever possible. In the Hepburn Wind project, this led to $2 million of construction costs being spent within the region.[297]

FIGURE 6.14 Aerial view of an offshore wind farm. Photo by Shaun Dakin from Unsplash.

100% renewable energy communities in Europe such as Samsø Island have led to a fall in electricity prices. When a region becomes a net exporter of renewable energy, the future electricity prices might not justify reinvestment in upgraded infrastructure.[298] Two mitigation strategies can be employed, firstly, invest in smaller scale turbines that are more affordable to repower when necessary. Secondly, invest part of the energy income into a trust dedicated to the repowering or decommissioning of the project at the end of the lifespan.

Micro hydro

Upper Yarra Community Power[299]

1. Jurisdiction/region

The Yarra River, Victoria, Australia.

2. Background of community participants

During the period from 1918 to the early 1940s, the community had a small hydro plant. The early Sanitarium Weet-Bix factory in Warburton, Victoria, ran on this hydropower plant. There were previously two deadlines on the creek the first one

FIGURE 6.15 An example of a micro hydro plant. Photo by Michael Telitsyn from Pexels.

was installed in 1932. The site of the new micro hydro was situated on the same site where hydro plant had existed 100 years earlier.

The engineering company Powershop helped get this project off the ground.

3. Technology used

Micro hydro is small-scale hydro that doesn't have a dam. It has a lower environmental cost due to the small intake.

The project site has about 100 kW/h when running 24 hours, seven days a week, although there are some seasonal high and low variations.

a. The viability equation:

$$Head\left(fall\ in\ meters\right)\times Water\left(L\ /\ second\right)\times 9.8\left(gravity\right)$$
$$\times Efficiency\left(assumed\ 70\%\ approximately\right) = Power\left(watts\right)$$

4. Funding model

$450,000 in grants was awarded through the New Energy Jobs Fund. A $450,000 loan was given by the parent company. $285,000 of shares were issued to raise capital. Some companies provided pro bono services such as Engineers Without Borders and Soft Loud House Architects.

5. Legal/governance model

Upper Yarra Community Enterprise (UPCE) is an unlisted public company limited by shares. 1.7 million shares were issued. There are 532 shareholders. The company was established in 1998 when it purchased a community bank franchise. A second capital raising in 2007 allowed the purchase of a second community bank franchise. Upper Yarra Community Power Pty Ltd is a private company established in 2018 and wholly owned by UPCE.

The hydro project required a lease over the subject land (the local golf club). They also needed a nonconsumptive water licence from Melbourne water. They had to organise a connection approval from Ausnet. Planning and building approval was obtained from the Yarra Ranges Council. Aboriginal Cultural Heritage Agreements were negotiated.

Cost-benefit analysis

6. Benefits

The lifespan of the hardware (technology) is approximately 100 years. This is much more sustainable than other renewable energy infrastructure.

7. Costs

Renewable energy projects are susceptible to changes in FiTs. In 2013, the project stalled when the FiT dropped from $0.21/kWh to 8 cents/kWh. The project then secured grants to continue. Additionally, the local council became a partner to assist with the planning approvals, grant applications and other legal obstacles such as biodiversity assessment, tree removal, cultural heritage and planning approvals.

FIGURE 6.16 A large solar farm in a rural field. Photo by Andreas Gücklhorn from Unsplash.

Earthworks were required to buffer the sound from neighbouring properties.

Planning took eight years, mostly due to the regulatory changes that occurred during the project, requiring it to be amended.

8. Learnings and recommendations

It is important to assess the viability of the project in terms of local governance. Where the approvals process is going to stall a project, the particular community might not be a viable location unless partnership with the local council is an option for the investor.

Network solutions for agricultural communities

Bomen Solar Farm, Wagga Wagga

1. Jurisdiction/region

Wagga Wagga is an agricultural city in New South Wales, Australia. Multiple solar farm projects have been approved in the city of Wagga Wagga. This case study is the Bomen Solar Farm which is located on approximately 250 ha of industrial mixed-use land.[300]

2. Background of community participants

This project started out as a small private company that was later bought by a large publicly listed entity, Spark Infrastructure Group. In 2021, Spark Infrastructure Group was acquired by Pika Pty Ltd and the company was delisted from the stock exchange. Pika Pty Ltd is indirectly owned by

> funds and/or investment vehicles managed and/or advised by Kohlberg Kravis Roberts & Co. L.P. and/or its affiliates, Ontario Teachers' Pension Plan Board and Public Sector Pension Investment Board. It acquired 100% of the issued securities in Spark Infrastructure Group by way of a creditors' scheme of arrangement, a trust scheme and related transactions.[301]

3. Technology used

Solar PV.

4. Funding models

The Bomen Solar Farm was funded partly under a ten year power purchase agreement with Westpac Bank through an agreement with the commercial electricity retailer Flow Power. Westpac will source 63 GWh of solar energy annually from the farm. Westpac has stated that it is through this PPA that it is able to underwrite the development of the new solar farm. Other PPAs with the Sydney Opera House and Molycorp Group will also help fund the project. Spark estimates revenue of around $13.5 million a year and a price of less than $60/MWh including the sale of the renewable energy certificates.[302]

5. Legal/governance model

The Bomen Solar Farm was owned by publicly listed entity, Spark Infrastructure Group while in the development phase. The company structure was made up of various trust entities, partnerships and holdings as well as partnerships with privatised grid operators. Income was derived through share dividends, loan repayments and trust distributions.[303]

Cost-benefit analysis

6. Benefits

This project creates employment opportunities in the agricultural region of Wagga Wagga. The Solar Farm saves on CO_2 emissions. It provides economic and environmental benefits for this agricultural region of New South Wales. It provides renewable energy for the local community as well as exports renewable energy to urban infrastructure such as the Sydney Opera House through a lucrative PPA.[304]

7. Costs

Spark Infrastructure partnered with several transmission networks in Victoria and South Australia as well as Transgrid. The farm powers approximately 36,000 homes as well as the Sydney Opera House and Westpac Bank. To create a large scale network solution such as this, partnerships with privatised transmission grids or public grids are necessary. This requires legal contract negotiation with large entities which could be a barrier for small communities or private entities.

8. Learnings and recommendations

To compete in the electricity export market, partnerships or relationships with network providers are beneficial. Large scale projects in agricultural regions with high solar can be profitable for public entities, funds and pension investments.

Terrain Solar[305]

1. Jurisdiction/region

Rural Australia.

2. Background of community participants

FIGURE 6.17 A large-scale solar plant integrated with a cattle farm. Photo by Dzmitry Palubiatka from iStock Photo, Getty Images.

Terrain Solar is building solar farms across rural Australia. For example, it has projects in Wagga Wagga, Junee, Corowa, Warwick, Molong, Marulan, Myrtle Creek, Monaro, Kingaroy and Moama.

3. Technology used

Large scale solar.

4. Funding model

The projects are largely funded by PPAs.[306] For example, the project in Wagga Wagga is in a ten year power purchase agreement with Coles alongside two of Terrain Solar's other solar farm developments, providing approximately 10% of Coles' energy Australia wide.

5. Legal/governance model

A private company that partners with landowners.

Cost-benefit analysis

6. Benefits

The company provides local jobs. The solar farms are generally integrated into agricultural land. For example, sheep and cattle farms.

7. Costs

It can take longer for an independent private company to get timely grid connection.

8. Learnings and recommendations

A private company underwritten with a PPA is a sensible and simple model for producing renewable energy. Additionally, partnering with agricultural landowners is a sensible solution to overcoming the tenure or zoning objection issues that could otherwise be faced. Partnering with transmission entities can expedite development.

Rural Europe

Varese Ligure[307]

1. Jurisdiction/region

Varese Ligure is an agricultural town in northern Italy, as depicted in Figures 6.18 and 6.19.

FIGURE 6.18 The historical town centre ofVarese Ligure, La Spezia, Liguria, Italy. Photo by Imen Chakir from Unsplash.

2. Background of community participants

One hundred per cent ofVarese Ligure's energy comes from renewables. The town covers 14,000 ha of mountainous forest and farms with a population of 2,400 people. There are 108 organic farms that produce almost all of the town's food. The region is broken up into 27 hamlets. It was the first European community to receive ISO 14001 certification in 1999. The region has won several awards for its renewable energy projects.

3. Technology used

A mix of wind, solar, biomass and small-scale hydropower make up the 100% renewable energy self-sufficiency. The region has four 46 m tall wind turbines positioned 1,100 m above sea level producing approximately 2 million kWh per year.

The local government funded solar panels on the roofs of the public buildings, the town hall, school, local heated swimming pool and public waste water treatment station. Solar is also installed in the town's only hotel.

Due to the agricultural industry, a lot of organic waste is produced. Waste from agriculture and factories is dried and compressed to produce pellets for fuel. This ensures the local forests are protected from being used as a fuel source.

FIGURE 6.19 Aerial view of the town of Varese Ligure, La Spezia, Liguria, Italy. Photo by Josè Maria Sava from Unsplash.

4. Funding model

The projects were initially funded through Federal and EU government grants and financially sustained long-term by a joint private partnership among an energy company (ARE Liguria SPA), public energy company (ACAM) and local government.

The funding is a mixture of private company funding and local government investment. The local government earns around €30,000 a year in surplus electricity generation. At the beginning of the town's rejuvenation, funding was received from the EU, and the Italian government. These funds went towards building renewable energy infrastructure and restoring the cultural heritage of the historical town centre.

Local community groups also raised funds and private investors contributed.

Additionally, 50 agricultural cooperatives have reported good profits from their own renewable energy systems.

The local government also receives an additional €350,000 in revenue from company tax on private renewable energy companies.

5. Legal/governance model

The wind turbines are installed, managed and maintained by the power company, ACAM and jointly owned by the local government and a private company, ARE Liguria SPA.

Cost-benefit analysis

6. Benefits

The goal of 100% renewables and 100% organic agriculture was formulated in the 1990s when the local population had shrunk to more than half of its original size, causing economic decline. The successful strategy was devised by the local government and Mayor to drive the agricultural sector, energy sector and promote tourism, thereby increasing the economic wealth for the local community. Additionally, social and cultural benefits of the preservation of the historical town centre were achieved as part of the project. "Acceptance and participation by locals ensured fluent implementation" and increased awareness of environmental issues among the villages.

7. Costs

The town Mayor's main challenges were the bureaucracy in the rural Italian villages. High connection costs to the national grid and sourcing funding were also challenges. It took one year for the approval from the Italian government for wind farm project and another year to be connected to the national grid. However, the project came in under budget at half of the original cost quoted by the main electricity supplier in Italy. The expensive main electricity supplier did not win the build contract.

8. Learnings and recommendations

Without the renewable energy projects, Varese Ligure would have continued to have been disconnected from the main electricity transmission lines. This is because of its remote location and low population. Within the space of ten years, 140 new jobs were created within the renewable energy sector. Not only have the local people stayed in their villages but also former residents have returned and new residents are relocating to the area, attracted by the clean air, environmental benefits and organic produce.

The electricity from the wind turbines alone has reduced carbon emissions by 8,000 tonnes. A combination of a variety of financial mechanisms for funding, support of the local council and the involvement of the people in the farming community has created a locally based model for 100% renewable energy and self-sufficient organic farming community that would otherwise have been disconnected from the electricity grid.

FIGURE 6.20 Indigenous Amazonian village. Photo by Atlantic-kid from iStock Photo, Getty Images.

Rural South America

Light at home[308]

1. Jurisdiction/region

In Latin America, there are 17 million people without electricity due to the isolated locations which have made it difficult to extend the grid to those areas. It has been recognised that financial mechanisms are needed to promote off-grid solutions with solar technology.[309]

Light at home has constructed solar off-grid systems in Latin America. This case study compares the systems built in Mexico and Peru.

2. Background of community participants

Four hundred small solar systems have been installed in the heart of the Amazon rainforest (Peru) near the Napo River, see Figure 6.20, to people who would otherwise have no electricity.

In Mexico, 7,500 households (30,100 users) have been supplied with solar energy in the Oaxaca region where grid expansion was not viable.

3. Technology used

This case study looks at off-grid solar PV for individual home systems.

4. Funding model

These projects are co-funded by a non-profit organisation Acciona.org and grants from the Multilateral Investment Fund and the Inter-American Development Bank. Funds are also obtained from the National Fund for Scientific and Technological Development and Technological Innovation (FONDECYT), The Mexican Agency for International Development Cooperation and the AECID.

In Mexico, the customers had to pay half of the cost of installation and equipment (US$150). Some customers accessed micro finance to fund the solar installation.

5. Legal/governance model

The projects are owned by the non-profit organisation and customers pay a monthly fee for the energy service which includes the daily supply of electricity. Micro franchises are given to local people who provide sales and bill collection services to the non-profit. They sell electricity but also appliances that are energy efficient.

Cost-benefit analysis

6. Benefits

In Mexico, 12,510 people have benefited from the programme. 11, 800 families have solar installations, there are 650 micro businesses such as grain mills, water pumps and electric fences. Sixty communities have received social benefits such as electric pumping for drinking water, school classroom lighting and electrified health centres. Sixty jobs with an increase in income of 20% or more were created in 1 year alone.

7. Costs

Micro franchising creates employment and income opportunities for local communities. However, while the goals of micro franchises are to create entrepreneurs, it was found that the majority of participants only engaged in billing and collection tasks and that some of the revenues were poorly designed which discouraged collection. The continuation of the infrastructure relies on continued grant funding.

8. Learnings and recommendations

The non-profit organisation structure attracted significant investment, grant money and support from funds and government agencies. Additionally, it was able to involve local community members in creating businesses and employment opportunities through micro franchising. Overall, it has been a successful programme in Latin America.

FIGURE 6.21 Indigenous Amazonian mother and child. Photo by FG Trade from iStock Photo, Getty Images.

Rural Africa

Elephant Energy[310]

1. Jurisdiction/region

Namibia, rural Africa.

2. Background of community participants

Participants in this program were off-grid customers from rural African communities. Namibians live in isolated remote farming homesteads or semi-nomadic villages.

3. Technology used

Solar powered lights and chargers.

4. Funding model

Elephant Energy was a non-profit organisation that relied on grants and donations. Some customers used funds generated from community-based hunting and tourism businesses to purchase solar products.

A rent-to-own system allowed people to pay small amounts towards ownership of their products. The rent-to-own model is recognised as an efficient non-profit distribution model but did not create a profit for investors. Grants and donations were relied upon for operational and oversight costs. It was also difficult to collect rents from people who couldn't afford the payments so a cloud-based operating system was used which switched off the electricity when the payments lapsed. Ultimately, this system was not sustainable in the long term and the payments declined during drought season impacting on the organisation's ability to continue to operate.

5. Legal/governance model

Elephant Energy was a non-profit organisation that distributed solar-powered lights (as shown in Figure 6.22) and chargers to rural communities that would otherwise be without electricity. They have distributed over 15,000 solar-powered products in Namibia. 28.5 million solar lights have been distributed in total in Africa by a variety of non-profit organisations.

FIGURE 6.22 An example of a solar light product. Photo by Nothing Ahead from Pexels.

Cost-benefit analysis

6. Benefits

The initial goal and benefit was for farmers to be able to use solar lights to keep elephants away from their crops. This not only protected crops but also helped in decreasing elephant hunting and poaching. Elephant tourism gradually replaced the poaching industry.

Following the success of this initial project, Elephant Energy started distributing other solar powered products. It was found that many additional community benefits came out of electrification, such as shops staying open at night, cell phone uptake to improve communication and access to information, children studying at night, people avoiding dangerous animals and snakes at night, women feeling safer and health clinics being built. Some of the stories from the local people include:

> I am very happy to use this product because now I'm not spending money to buy some needs like candles, matches and pay to charge cell phone each week. Now everything is possible.
>
> *Susana Ndawedwa – Ohangwena, Ohangwena Region*

> I am very proud for this solar product. The children are performing well now at school because of that.
>
> *Teopolina Daniel – Ongha, Ohangwena Region*

> It helped us to check our chickens during night when the wild cat wanted to eat our chickens. It also helped us to see where you walk in order to check for snakes.
>
> *Lucious Mafwila – Kasika, Caprivi Region*

7. Costs

A challenge of supplying products in Namibia was the shipping costs to the isolated regions and the risk of theft during transport.

Another major challenge was the process and stipulations associated with the grants needed to fund the non-profit. In particular, the non-profit had to cover operating costs between grant periods in order to retain staff and operations in Africa.

The free products were handed out by local volunteers who gave preference to their family and friends and did not share with the wider population, so the non-profit charity model was replaced with a subsidized sale model. Due to the low profits, the organisation realised it would not be sustainable in the long term, so an energy shop was opened in the town centre to sell electrical products. That business as well as donations and volunteer staff kept the non-profit running.

Elephant Energy was asked to provide charitable donations and grant-based solar supply to flood victims but found that the charitable donation model was not sustainable and that the time spent on those donations would have been more productively spent on building its rural distribution business. After six month of disaster relief work, it took two years for the non-profit to rebuild its staff capacity and business capacity to provide solar products to the flood effected areas. As the grant-based model had onerous policy requirements and stipulations, it did not allow Elephant Energy to bring its energy shop and energy product distribution network into the flood-effected region. This meant that the local people could not repair, maintain or replace their solar and electrical products. The opportunity to expand into those unelectrified villages was lost. After that experience, Elephant Energy focused on grants that supported its entrepreneur-focused mission.

A further challenge was the role in addressing warranty issues with customers and suppliers over faulty products and products that could not withstand the conditions in the African environment. Due to long delays, in shipping and supply shortages, a shipment of defective products caused stock outages for lengthy periods of time which jeopardised the business and the sales.

8. Learnings and recommendations

Elephant Energy recognises that its success was due to its "deep connection with the country and its people." The non-profit chose to focus on one million unelectrified Namibians rather than on more lucrative population centres in Kenya or India because of the life-changing social benefits to the isolated communities.

It also recognises that not having to rely on the grant process for funds would have resulted in,

> thousands more sales and fewer sleepless nights. Elephant Energy, like many other social ventures, struggles with the demands of grantors and investors that do not understand the challenges of working in Africa with innovative new business models and technologies. Elephant Energy now takes the time to consider the experience and expectations of donors before accepting their support, despite the organization's constant need for funds to support new staff and inventory ....

Elephant Energy also recognises that the constant effort to raise funds from donors and grants distracted from the goal of forming a sustainable energy network.

Large investors were not attracted to a new non-profit with no track record and so small donations from friends and family were relied upon. The organisation was not self-sufficient economically and relied on donations and grants to continue operation. It is currently not operating.

FIGURE 6.23 Without electricity clothes are washed by hand by women an children. An African girl carries washing she has done by hand on her head. Photo by Richmond Osei from Unsplash.

Summary

In summary, many of the benefits of community renewable energy projects are social community benefits such as employment and training opportunities or environmental benefits, such as uptake of renewable energy or practical social benefits such as the ability to wash clothes and refrigerate food and medicine. As shown in Figure 6.23 without electricity, clothes are washed by hand by women and children.

Nevertheless innovative funding and governance structures can ensure that communities not only benefit from a social perspective but also benefit financially. The next chapter will review the case study findings in relation to the most successful structures for attracting investment and accessing finance as well as the most sustainable technology choices.

Notes

1 Centre for Appropriate technology limited, *Bushlight Archive*, at https://cfat.org.au/bushlight-archive/ accessed on the 4th of April 2019; Buergelt, P., Maypilama, E., Mcphee, J., Dhurrkay, G., Nirrpuranydji, S., Mänydjurrpuy, S., …, Moss, S. (2017). Housing and Overcrowding in Remote Indigenous Communities: Impacts and Solutions from a Holistic Perspective. In Energy Procedia (Vol. 121, pp. 270–277). Elsevier Ltd.

2 See, for example, Buergelt, P., Maypilama, E., Mcphee, J., Dhurrkay, G., Nirrpuranydji, S., Mänydjurrpuy, S., …, Moss, S. (2017). Housing and Overcrowding in Remote Indigenous Communities: Impacts and Solutions from a Holistic Perspective. In Energy Procedia (Vol. 121, pp. 270–277). Elsevier Ltd.

3 For instance, in remote Australian communities supply is usually provided by the power card system or diesel generator micro grids. When power cards run out of credit, the power supply to the house is cut off. See for example, Buergelt, P., Maypilama, E., Mcphee, J., Dhurrkay, G., Nirrpuranydji, S., Mänydjurrpuy, S., …, Moss, S. (2017). Housing and Overcrowding in Remote Indigenous Communities: Impacts and Solutions from a Holistic Perspective. In Energy Procedia (Vol. 121, pp. 270–277). Elsevier Ltd.
4 Sovacool 2014. *Energy Research & Social Science*, 14.
5 Ibid.
6 Ibid.
7 Centre for Appropriate Technology Limited 2016, 33.
8 Buergelt 2017, 266–267; 273–274.
9 Sovacool 2014. *Energy Research & Social Science*. From Masud, J., Sharan, D., Lohani, B.N. (2007). Energy for All: Addressing the Energy, Environment and Poverty Nexus in Asia. Manila: Asian Development Bank.
10 Sovacool 2014. *Energy Research & Social Science*, 14.
11 Ibid. 25.
12 Ibid. 13.
13 Goetzmann 2017, Ch. 17.
14 Ibid.
15 Ibid.
16 Ibid.
17 Ibid.
18 Ibid.
19 Ibid.
20 Ibid.
21 Centre for Appropriate Technology Limited 2016, 8, 59.
22 Centre for Appropriate technology limited 2014.
23 Northern Territory Government, *Projects*.
24 Bulletpoint 2012.
25 The Queensland Government, Department of Natural Resources, Mines and Energy. "Solar for Remote Communities", 2020.
26 SBS news 2020.
27 Australian Government, "Funding to support studies into improved energy supply arrangements".
28 Vorrath 2020.
29 EWB challenge 2019.
30 Energy Queensland 2018, 51.
31 Ibid.
32 Ibid.
33

> Resource agencies have been operating for many years. A manager is employed and funded through the NT Government and the Aboriginals Benefit Account and they live in the community. The resource agency's business activities include maintaining houses, managing municipal services (power, water, sewerage), undertaking civil construction works, delivering contracts for regional council, providing Centrelink services, providing internet access for the community and delivering other government-funded infrastructure projects.
>
> Centre for Appropriate Technology Limited, *The Northern Territory Homelands and Outstations Assets and Access Review*, Final Report 2016.8, 59.

34 Centre for Appropriate Technology Limited 2016, 8.
35 Ibid. 62.
36 Centre for Appropriate Technology (CfAT) at www.cfat.org.au
37 Ibid
38 Ibid.
39 Coalition for Community Energy 2017.
40 Grant et al. 2018, 120; Martin 2016, 15.
41 Nettheim 2000.
42 *Yarimirr and Others v Northern Territory and Others* (1998) 156 ALR 370; see also Langton et al. 2006.
43 Keim and Reidy 2015.
44 In December 2010, the Federal Court made a determination recognising the native title rights and interests of the Waanyi Peoples. The determination covers 1,730,081 ha and recognises exclusive possession over the Bidunggu Land Trust area and non-exclusive possession over a number of pastoral properties, reserves and Boodjamulla (Lawn Hill) National Park.
45 Aboriginal Land Tribunal, Annual Report, 2018–2019.
46 Environmental Defenders Office 2019.
47 Department of Natural Resources and Mines 2020. "Leasing Aboriginal Deed of Grant in Trust Land".
48 *Wigness & Ors v Kingham, President of the Land Court of Qld & Ors* [2018] QSC 20.
49 Department of Natural Resources and Mines 2020. "Leasing Aboriginal Deed of Grant in Trust Land".
50 CAT Homelands and Assets Report 2016, 36.
51 Ibid. 36.
52 CfAT. www.cfat.org.au
53 Bushlight 2014. Case Study 35.
54 Ibid.
55 Ibid.
56 Bushlight 2014. Case Study 7.
57 Ibid.
58 Bushlight 2014. Case Study 34.
59 Ibid.
60 Ibid.
61 Ibid.
62 Bushlight 2014. Case Study 33.
63 Ibid.
64 Ibid.
65 Ibid.
66 Ibid.
67 Ibid.
68 Ibid.
69 Bushlight 2014. Case Study 32.
70 Ibid.
71 Bushlight 2014. Case Study 31.
72 Ibid.
73 Bushlight 2014. Case Study 25.
74 Ibid.
75 Bushlight 2014. Case Study 22.

76 Ibid.
77 Ekistica 2020.
78 Ibid.
79 Desert Knowledge Australia 2019, 24. CfAT Annual Report, 2017–2018.
80 CfAT. www.cfat.org.au
81 Desert Knowledge Australia 2019, 24.
82 Ibid.
83 Ibid. 19.
84 CfAT Annual Report, 2017–2018.
85 Desert Knowledge Australia 2019, 24.
86 Ibid.
87 Ekistica 2020.
88 CfAT Annual Report, 2017–2018, 14.
89 Ibid.
90 CfAT Annual Report, 2017–2018, 13–15.
91 Tran 2016.
92 CfAT Annual Report, 2017–2018, 13–15.
93 Nehme 2014.
94 Ibid.
95 Ibid.
96 Australian Government, Office of the Registrar of Indigenous Corporations 2010.
97 Nehme 2014.
98 See Part 6-4 *CATSI Act 2006*; Australian Government, Office of the Registrar of Indigenous Corporations, 2010.
99 Ibid.
100 Ibid.
101 Goetzmann 2017. 13.
102 CAT Homelands and Assets Report, 2016.
103 Ekistica 2020, 7.
104 Centre for Appropriate Technology (CfAT) 2000, 5–6.
105 Ibid.
106 Bushlight 2014. Case Study 15.
107 Bushlight 2014. Case Study 11.
108 CAT Homelands and Assets Report, 2016, 62.
109 National Indigenous Australians Agency.
110 Solar Hybrids.
111 Johns 2011, 255.
112 Sekisui House 2018.
113 Australian Government Austrade 2017.
114 Ibid.
115 Ibid.
116 Ibid,
117 Droege 2009, Ch. 6; Australian Government Austrade 2017.
118 Droege 2009, Ch. 6.
119 Matich 2019.
120 Australian Government Austrade 2017.
121 Ibid.
122 Ibid.
123 Ibid.

124 Ibid.
125 Ibid.
126 Ibid.
127 Droege 2009, Ch. 6; Australian Government Austrade 2017.
128 Ibid.
129 Ibid.
130 Ibid.
131 Yunus and Weber 2017, 64.
132 Ibid.
133 Ibid.
134 Ibid. 13.
135 Keim 2018. "A World of Three Zeroes".
136 Yunus and Weber 2017, 8.
137 Meinzen-Dick 2019.
138

> The Social Success Note offers a new twist on this approach. It involves teamwork among three participants: a social business, an investor, and a philanthropic donor, such as a foundation. The investor provides funding in the form of a loan to the social business to pursue a particular, well-defined social goal— building homes for a certain number of homeless people, for example, or extending health insurance to a certain number of families. The social business is responsible for repaying the loan. But if it achieves the predetermined goal by an agreed deadline, the philanthropic donor will add on an impact payment to the investor. As I noted in an article in Bloomberg View, the Social Success Note creates a win-win-Investors receive a risk-adjusted commercial return, thanks to the impact payment; foundations achieve far greater leverage for their philanthropic dollars while achieving a desired social outcome; and social businesses receive access to low-cost capital, allowing them to focus on improving the world without the pressure of offering market-rate financial returns.
>
> *From, Yunus and Weber 2017*

139 Yunus and Weber 2017, 148–149.
140 Ibid.
141 Ibid.
142 Ibid.
143 Ibid.
144 Ibid.
145 Ibid.
146 Ibid.
147 Ibid.
148 Ibid
149 Ibid.
150 Ibid.
151 Ibid.
152 Reserve Bank of India 2020.
153 Yunus and Weber 2017, 148–149.
154 Rahman 2001, 147–152.
155 Ibid.
156 Ibid.
157 Ibid.

158 Ibid.
159 Ibid.
160 Ibid.
161 Ibid. 64.
162 Helm 2015, 186.
163 Ibid.
164 Ibid.
165 Yunus and Weber 2017, 64.
166 Lou et al. 2013, 283–294.
167 Australian Renewable Energy Agency 2019. "Biogas Opportunities for Australia".
168 Riesz and Elliston 2016.
169 Homebiogas 2020.
170 Australian Renewable Energy Agency 2019. "Biogas Opportunities for Australia".
171 Sovacool et al. 2014. *Energy security, equality and justice*, 98.
172 Business Queensland 2016.
173 Ibid.
174 Ibid.
175 Sovacool 2012. *Energy & Environmental Science*, 9157–9162.
176 Ibid.
177 Gibson et al. 2014, 26–35.
178 Australian Renewable Energy Agency 2019. "Biogas Opportunities for Australia".
179 Ibid.
180 Sovacool 2012. *Energy & Environmental Science*, 9157–9162.
181 Ibid.
182 Ibid.
183 Droege 2009, Ch. 6.
184 Ibid.
185 Ibid.
186 Ibid.
187 Ibid.
188 Ibid.
189 Business Queensland 2016.
190 Ibid.
191 Ibid.
192 Lou et al. 2013, 292.
193 Droege 2009, Ch. 6.
194 Andhina and Purwanto 2018, 73.
195 Riesz and Elliston 2016.
196 National Farmers' Federation Ltd.
197 Ibid.
198 Vidot, A, ABC Rural News.
199 Ibid.
200 Ibid.
201 Fitzgerald et al., ABC Rural News.
202 Ibid.
203 Ibid.
204 Ibid.
205 The Queensland University of Technology, newsletter (3 August 2020) "QUT PV solar seven days".
206 The University of Queensland Sustainability Office.

207 Coalition for Community Energy 2017. *Small-Scale Community Solar Guide*, v.2 .21.
208 Frontier Impact Group 2018, 61.
209 Victorian State Government. Coalition for Community Energy 2017. *Small-Scale Community Solar Guide*, v.2 .21.
210 Ibid.
211 Ibid.
212 Ibid.
213 Ibid.
214 Coalition for Community Energy 2017. *Small-Scale Community Solar Guide*, v.2 .21.
215 Frontier Impact Group 2018, 61.
216 Coalition for Community Energy 2017. *Small-Scale Community Solar Guide*, v.2 .21.
217 Ibid. 22.
218 Martin 2016, 41.
219 Hicks and Ison 2018, 530.
220 Ibid.
221 Citizens Own Renewable Energy Network Australia.
222 Coalition for Community Energy 2017. *Small-Scale Community Solar Guide*, v.2 .21.
223 Ibid.
224 Ibid.
225 Ibid.
226 Coalition for Community Energy, "C4ce webinar 6 2018 – revolving funds".
227 Ibid.
228 Ibid.
229 Ibid.
230 Ibid.
231 Ibid.
232 Citizens Own Renewable Energy Network Australia.
233 Coalition for Community Energy, "C4ce webinar 6 2018 – revolving funds".
234 Ibid.
235 Coalition for Community Energy 2017. *Small-Scale Community Solar Guide*, v.2 .21.
236 Ibid.
237 Ibid.
238 Ibid.
239 Ibid.
240 Ibid.
241 Ibid.
242 Ibid.
243 Ibid.
244 Coalition for Community Energy, "C4ce webinar 6 2018 – revolving funds".
245 Ibid.
246 City of Darebin.
247 Indigenous Business Australia, Homepage.
248 Australian Government, National Indigenous Australian Agency.
249 Coalition for Community Energy 2017. *Small-Scale Community Solar Guide*, v.2 .21.
250 Repower Shoalhaven.
251 Coalition for Community Energy 2017. *Small-Scale Community Solar Guide*, v.2 .21.
252 Common Capital 2015, 13.
253 Coalition for Community Energy 2017. *Small-Scale Community Solar Guide*, v.2. 15.
254 Frontier Impact Group 2018, 115.

255 Ibid. 133.
256 Coalition for Community Energy 2017. *Small-Scale Community Solar Guide*, v.2. 17.
257 Ibid. 15.
258 Ibid. 24.
259 Ibid. 17.
260 Ibid. 15.
261 Ibid. 18.
262 Frontier Impact Group 2018, 119.
263 Ibid.
264 Coalition for Community Energy 2017. *Small-Scale Community Solar Guide*, v.2. 23.
265 Frontier Impact Group 2018, 85.
266 Coalition for Community Energy 2017. *Small-Scale Community Solar Guide*, v.2. 18.
267 Ibid. 16.
268 Ibid. 18.
269 Ibid. 18.
270 Frontier Impact Group 2018, 115.
271 Coalition for Community Energy 2017. *Small-Scale Community Solar Guide*, v.2 .21.
272 Hicks 2018. *Community Power*, 93.
273 Droege 2009, Ch. 6.
274 Hicks 2018. *Community Power*, 93.
275 Hicks 2018. *Community Power*, 92–95.
276 Droege 2009, Ch. 6.
277 Hicks 2018. *Community Power*, 110.
278 Ibid. 163.
279 Ibid.
280 Droege 2009, Ch. 6.
281 Ibid.
282 Hicks 2018. *Community Power*, 163.
283 Ibid. 194.
284 Ibid. 60.
285 Ibid. 179.
286 Ibid. 185.
287 Droege 2009, Ch. 6.
288 Hicks 2018. *Community Power*, 168.
289 Ibid. 168.
290 Hicks 2018. *Community Power*, 166.
291 Droege 2009, Ch. 6.
292 Hicks 2018. *Community Power*, 120.
293 Ibid. 145.
294 Lane 2017. "Investigating integrated systems of European zero net energy villages", the Winston Churchill Memorial Trust, 90.
295 Victorian State Government (2015), Guide to Community-Owned Renewable Energy for Victorians, www.Ecodev.Vic.Gov.Au, accessed 10 April 2020.
296 Hicks 2018. *Community Power*, 157.
297 Ibid. 157.
298 Lane 2017.
299 Coalition for Community Energy 2018. Webinar 7.
300 Bomen Solar Farm.
301 Spark Infrastructure, "Investor Centre".

302 Ibid.
303 Ibid.
304 Parkinson 2020.
305 Terrain Solar.
306 Ibid.
307 Droege 2009, Ch. 6.
308 Eras-Almeida et al. 2019, 7139.
309 Eras-Almeida et al. 2019, 7139.
310 Guruswamy and Neville 2016, Ch. 17.

7

ATTRACTING INVESTMENT

Introduction

Access to finance and attracting investment funds is the key issue for remote energy infrastructure. Much of the funding for remote community infrastructure investment is attracted through public benefit goals. Governments rightly view it as their joint responsibility to provide initial funding for energy infrastructure to remote communities to provide increased standards of living and remove energy access disparity. Internationally, funding is often provided by the World Bank or, more recently, by large institutional investors, such as superannuation funds. Non-profits and NGOs are also involved. There are usually a variety of grants available in at a national level.[1]

Funding for remote community energy infrastructure has increased significantly as various agencies and organisations compete for control over energy infrastructure in remote regions. In the past, this investment has not been sustained with ongoing maintenance and upgrades to the energy systems. However, the case study results demonstrate that innovative funding models can attract investment and sustain remote micro grid energy infrastructure long term without the need for ongoing public investment or donations. This creates an effective balance between private investment, community participation and government investment. The case studies demonstrate that sustainable energy projects attract population to remote regions and reinvigorate local economies and that the subsequent economic growth causes improved standards of living for those communities.

Funding arrangement selection

Below is a summary list of successful funding arrangements that have attracted investment in community renewable energy infrastructure projects and have worked well with the governance structures, as analysed in the case studies.

DOI: 10.4324/9781003324669-7

Firstly, initial capital investment needed to be attracted to the project. In the case studies, a variety of capital was successfully attracted to renewable energy projects as listed below:

1. Private investment
2. Trust income
3. Public funding/grants (Local, State, Territory, Federal and International as well a specific bodies such as the Indigenous Business Association)
4. Donations/charity
5. Issuing shares
6. Loans, secured
7. Loans, unsecured
8. Loans, interest payable
9. Loans, no interest payable
10. power purchase agreements
11. Revolving funds
12. Dividends from subsidiaries
13. Joint venture partner investors
14. Microcredit (no structure or contracts)

Secondly, to sustain a project ongoing income or the ability to raise capital was necessary to the success. This ensured that maintenance, upgrades and decommissioning was feasible. It also served as an attractive carrot for potential investors in the enterprise design phase of the project. The successful income streams that attracted investment in the case study renewable energy projects were: Dividends, FiTs, PPAs, the sale and/or trade of renewable energy certificates and trust income sustained by income from an industry in the community. Income producing industries noted in the case studies included:

1. Tourist enterprises;
2. Market gardens;
3. Maintenance and construction;
4. Grocery stores;
5. Renewable energy education centres;
6. Renewable energy training centres;
7. Farming;
8. Sale of electrical goods and services; and
9. Rent from properties where renewable energy infrastructure was built.

Choice of legal structure

The choice of legal structure or governance structure was crucial to the success of the renewable energy project and usually determined the type and quantity of investors that were attracted to the project. The following legal structures were successful in attracting investment.

Aboriginal and Torres Strait Islander corporations

An Australian Aboriginal and Torres Strait Islander controlled corporation is formed under *The Corporations (Aboriginal and Torres Strait Islander) Act* 2006 (Cth) (CATSI Act). It is usually also an incorporated not-for-profit company.

Aboriginal and Torres Strait Islander corporations attract funding more readily than communities that are not incorporated. CfAT was awarded $1,115,509 last year in Commonwealth grants through its efforts as a registered Aboriginal and Torres Strait Islander corporation and not-for-profit company.[2]

The benefits of an Aboriginal and Torres Strait Islander Corporate structure include the empowerment of Indigenous communities in their ability to negotiate with government and other private interests. Incorporation offers a legal entity under which a community can conduct business. It has been viewed as an "intercultural phenomena" because although corporate structures are not traditional Indigenous ways of engaging, they can be used by Indigenous communities to engage with the wider business community.[3]

Trust funds

Trust funds have been used successfully by communities to prepare for the future maintenance and upgrades of the renewable energy systems. Some communities such as the Mabunji people have been forward thinking in preparing for the time when their grant ends. They put aside money into a trust fund so that the systems can be repaired or upgraded in the future. This has worked well in their community, allowing for upgrades and maintenance to battery storage systems that other similar communities were unable to afford following the end of their grant.[4]

Another novel use of the trust fund is for monies collected from community industry such as tourism or mining. For example, the Ulpanyali people put the money collected from Uluru tourism into a trust fund and then used their trust fund money to build solar infrastructure.[5] The community states that they have saved money on diesel for the backup generator as well as saving on the drive to Alice Springs for fresh food as they now have reliable fridges and freezers.[6]

Cooperative with member investors and/or a certified B-Corporation (B-Corp)

A Cooperative with member investors and/or a certified B-Corporation (B-Corp) is suitable for communities that want local control and do not want to be limited in the number of local investors.[7] Other structures such as a proprietary company do not allow all of the community members to invest because they have a limit on the number of investors. While Aboriginal and Torres Strait Islander corporations do not have a limit on the number of investors, other non-indigenous rural or agricultural communities who form a proprietary company in Australia have a limit of 20 investors every year and 50 in total. If a community project is large, there may be many more community investors interested in participating.[8]

A cooperative does not need to prioritise the profit of the shareholders but can set community benefits as a priority. This is especially useful if the community wants to set the Sustainable Development Proposition (SDP) as a priority. Sustainability of the system for the benefit of the community can be prioritised over and above a purely profits-based model.

The difficulty with the cooperative structure is that it cannot be used to sell electricity to the members. By contrast, a fishing or grocery cooperative means that an active membership is sufficient to enable a person to buy goods from the shop and gain a corresponding benefit through reduced cost of the goods.[9] It remains to be seen as to whether this anomaly in the regulations is rectified.

Hepburn Wind found that these kinds of cooperative structures attracted investment from locals in communities that were wanting local ownership and control and wanted to restrict outside developers from entering their rural locality. This resulted in a large quantity of small investment amounts as opposed to a limited number of large investment amounts that were attracted to private and public company structures in other case studies.

Proprietary company and council partnership

Proprietary Company and Council partnership structures consist of a private company that is funded by an unsecured or secured loan from the local council. The aim is to build infrastructure or provide income for the local council. Community investors can buy shares in the private company. Fully franked dividends are returned to the shareholders. This attracted community investment, large public grants and private company investment in the case studies.

Investor's initial loan capital is also returned at the end of the project. The investor's rate of return depends on the rate of interest on the loan and the repayment terms (interest only or principal and interest). The community investors are governed by the private company (SPV) rules and requirements under legislation.[10]

The proprietary company and council partnership structure attracts community investors who are interested in the profits-based model and do not want to be concerned with maintenance or refurbishment of infrastructure. It requires a good relationship with the local council.[11]

The return on investment is linked to the loan repayments by the council rather than the price of electricity under a PPA, as in other comparative private company models. This could be seen as an advantage or disadvantage depending on the price of electricity and the loan terms.[12]

The original investors loan money to the council, and the council builds, owns and operates the infrastructure. The original investors receive interest on their capital over a term until the loan is paid off. The council, on the other hand, receives the income from the operation of the infrastructure. Shareholder limits for private companies reduce the amount of investors.

Subsidiaries

Corporations can create subsidiaries. The Aboriginal and Torres Strait Islander corporation, CfAT, created the subsidiary Ekistica to undertake engineering works for its Indigenous community solar projects. In 2019, the subsidiary paid a fully franked dividend of $120,000 to CfAT.[13]

Subsidiaries are useful when a company expands into other industries or specialties. They can be profitable to the parent company. The profitability and legal governance advantages of creating a separate entity attracted investment.

Statutory corporation

DKA is a statutory corporation of the Northern Territory. Working in collaboration with the Australian Renewable Energy Agency (ARENA), a memorandum of understanding has been established with this federally funded agency. The mutual aims are to support local non-network renewable energy systems that can be later connected to the Alice Springs Future Grid Project. States and Territories use statutory corporations to attract funding from the Federal government or to negotiate investment from the private sector. The DKA has attracted large sums through this MOU contractual structure.

Not-for-profit

Not-for-profit SPVs are governed by legislation.[14] The legal and accounting requirements are onerous and the transaction costs are high. Some examples such as Elephant Energy in Africa were heavily grant and donation dependent and did not attract private investment so were not sustainable long term. However, organisations such as ClearSky in rural Australia has attracted investment but still relies on volunteers and grants to cover capital, operational and transaction costs.[15]

Private company

Private companies are successful where community involvement is not essential. The investors in Australia are limited to 20 in any year and a maximum of 50. The return on investment is the key factor in attracting investors with this model. A proven track record of successful projects will attract investment in subsequent projects. A Power Purchase Agreement demonstrating the return on capital and operational investment cost can assist in attracting investment.

The solar PV infrastructure is usually retained by the company, and, therefore, the company has the obligation for maintenance, upgrades and decommissioning.

The costs of the following need to be taken into account: fundraising (equity and debt), tax, other legal costs, compliance and auditing requirements, insurance, operating costs, structure set-up and operation.[16]

Saving on the retail cost of electricity would not be enough to warrant this governance structure for a small PV solar system installation as the set-up costs could be expensive.

If there is no PPA to rely on for income, a system that is less than 100 kW has less risk on investment, keeping in mind that much of the income would then be coming from trading the small-scale renewable energy certificates.[17]

Where the project is an off-grid non-network solution and is selling electricity directly to the host site, the project will be much more viable as the host site will be saving retail electricity costs, and the project will be selling electricity for more than the wholesale price, making the return on investment more attractive.[18]

Unlisted public company

Upper Yarra Community Enterprise (UYCE) is an unlisted public company limited by shares. A total of 1.7 million shares were issued. There are 532 shareholders. The company was established in 1998 when it purchased a community bank franchise. A second capital raising in 2007 allowed the purchase of a second community bank franchise. Upper Yarra Community Power Pty Ltd is a private company established in 2018 and wholly owned by UYCE.

Hydropower is expensive to fund. It requires extensive engineering work, legal work and property leases/ownership as well as water licenses. This governance model allows shares to be issued for fundraising. In addition, the company applied for grants and was successful. The company was also able to secure a loan from the parent company. This structure was also successful in the profitable Bomen case study.

The advantage of a company structure is that capital can be raised, this is important not only at the construction phase of a project but also where the project needs to raise capital for maintenance, upgrades or decommissioning.

This structure has high transaction costs. It is suitable for an experienced community organisation with significant resources and professional expertise; in this case, the community bank suited those prerequisites. Therefore, this structure is useful for attracting investment in projects that are high cost, high return.

Innovative governance and financing structures

What has become apparent from the case studies is that renewable energy infrastructure projects reliant solely on grant or donation funding alone are not technologically sustainable (because they cannot keep up with ongoing maintenance, upgrades or decommissioning), and they are not financially sustainable for the ongoing supply of reliable electricity to the community without subsequent donations.

Innovative financing mechanisms are needed, which combine initial grant money, donations or private capital investment with a funding structure that can

FIGURE 7.1 Wind turbines in an agricultural field. Photo by Tom Fisk from Pexels.

ensure the ongoing sustainability of the renewable energy infrastructure from a technological, social, environmental and economically sustainable perspective.

In addition to the funding and governance structures identified in the case studies and outlined above, highlighted below are some hybrid innovative structures that could apply in the context of certain remote off-grid communities to attract funds:

1. Initial grant, sustained long term with income from the electricity generation that is placed into a trust fund.

   Income producing examples include:
   a. Sale of electricity through;
      i. FiTs; or
      ii. PPAs; or
      iii. Community energy retailers which have a higher than usual FiT for residential community organisations; and/or
   b. The sale and/or trade of renewable energy certificates.
2. Initial grant, governed by a trust fund structure, sustained by income from an industry in the community. Industries noted in the case studies include:
   a. Tourist enterprises;
   b. Market gardens;
   c. Maintenance and construction;
   d. Grocery stores;
   e. Renewable energy education centres;

   f. Renewable energy training centres;
   g. Farming;
   h. Electronics shops; and
   i. Rent from properties where renewable energy infrastructure is built.
3. Initial grants or donations sustained by a revolving fund governed under one of the following structures:
   a. A proprietary company;
   b. Trust fund;
   c. Trust; or
   d. Aboriginal and Torres Strait Islander corporation.
4. An Aboriginal and Torres Strait Islander revolving fund:

An Aboriginal and Torres Strait Islander corporation could borrow from or establish their own revolving fund specific for Indigenous communities.

The primary benefits of this model are that the Aboriginal and Torres Strait Islander corporation could use the model to build infrastructure across a number of communities and after the initial grants or donations, the fund could sustain itself in a financially Sustainable System without relying on further donations or grants.

Alternatively, an Aboriginal and Torres Strait Islander corporation could set up a revolving fund by partnering with the Indigenous Business Association[19] or National Indigenous Australians Agency[20] if assistance with governance, initial grant money and technical expertise is required.

5. MOUs:

Where a loan structure is unsuitable, an MOU for repayments can be used as seen in some case studies.

6. Partnerships and joint ventures:

A similar structure to the council and private company investment partnership found in the Lismore Shire Council case study and the rural Italy case study could be adapted to apply to a partnership between a private company and an Aboriginal Land Council or other Trustee with the project funded by an unsecured loan.

A primary consideration is the workforce. Partnerships and joint ventures ensure that localised services are met with local training and employment to negate the issue of a transitory workforce for maintenance. This has been addressed by CfAT through joint ventures with employment agencies, universities and industry colleges. The Bushlight program also gave some ad hoc training to individuals as needed until the grant money ended.[21]

Another successful model was the Bushlight remote maintenance service, where community members were given initial training and a user manual and were given access to a telephone maintenance hotline.[22]

Community Service Agreements (CSAs) should be well thought out and the parties held accountable. The CSA is an agreement between the community, its support or resource agency, the agency funding maintenance of essential services and the installer, where each party agrees to work together, in a spirit of cooperation, to maintain and sustain the energy services. The CSA must clearly articulate the roles and responsibilities of each party as well as describing maintenance and repair and arrangements.[23]

Choice of technology

Solar PV

Many of the case studies reviewed found that solar PV was the most successful renewable energy source for a remote community that was not connected to the grid.

The Pacific Renewable Energy Project trialled small-scale hydro, wind, geothermal, biodiesel and solar PV, and the most successful technology installed was small-scale solar PV.[24]

The Bushlight case studies, which involved 30 different communities likewise, concluded that solar PV was the most effective technology. Solar PV that had a backup diesel generator was the most reliable system from the Bushlight case studies.

The payback period for solar PV was different depending on the case study and the FiTs. Some case studies had a two to three year payback time. Others were slightly longer. The time is shorter for a larger system. It is also shorter for a system that is mostly used in the daytime.[25]

Biofuel and waste plastic oils

There are four types of energy systems that CfAT reports as being used to deliver electricity to remote Indigenous communities.

- Diesel/petrol generators (micro grids or individual systems);
- Solar PV (micro grids or individual systems);
- The traditional grid; and
- Hybrid (combination of solar power and back-up diesel generators).[26]

Until small-scale batteries reduce in cost, the recommendation for investment following on from this case study is investment in innovative projects where hybrid systems powered by solar with back-up waste plastic oil or biodiesel generator. Such innovations would ensure affordable electricity with the additional benefit of reliability provided by the backup generator. If the backup generator is powered by biodiesel or waste plastic oil, the result is a much more sustainable energy system.

The sustainability of biodiesel depends on the feedstock for the biofuel being free and renewable (such as waste oil from restaurants). This means that the fuel does not need to be specially grown in a crop.

Wind

There are many successful micro grid windfarms throughout the world. This research looked at the Hepburn Wind Farm and Samsø case studies as they are among the most successful rural small-scale wind farms. They are also both community projects.[27]

As with other renewable energy projects, the choice of property is key to a successful wind project. For example, the property should be checked for tenure, planning impediments or endangered species; access to suitable grid connection options, construction access, and any visual and noise impacts.[28]

Wind maps should always be consulted before selecting a host site. Wind maps can be used to analyse the potential of the wind energy on the subject site.

Biogas

Small-scale

Small-scale biogas plants used to fuel household stoves have been funded across India and China through low interest loans, grants and microcredit.

This technology could also be used by households in regional Indigenous communities where solar and wind are not viable options and where wood stoves or open fires are causing health and environmental hazards to the household, so long as sufficient installation warranties and maintenance training is in place, the OPEX is kept to a minimum, as demonstrated by the successful case studies in Northern China.

Mid-scale

Mid-sized plants for agricultural communities will attract investment where additional sources of revenue are available to offset the CAPEX and OPEX. For instance, waste treatment, sale of energy and sale of processed digestate as a fertiliser.

Overall, biogas is useful as a renewable energy source where large quantities of free biomass are available for processing on the subject site, such as a landfill or human or animal effluent. It is also useful as an alternative renewable energy source for national parks where tree coverage prevents solar capture, or any subject site where trees are to be preserved and there is too much shade to collect sufficient solar energy.

High quality maintenance is necessary to prevent ecological harm from leakage of slurry or gas. Strict planning legislation is in place in western countries to ensure plants are built to high standards. Additionally, consumer law should protect communities where installation warranties are provided.

Hydro

Micro hydro is small-scale hydro that does not have a dam. It has a lower environmental cost due to the small intake of water.[29] The lifespan of the hardware (technology) is approximately 100 years, which is much more sustainable than other renewable energy infrastructure.[30]

Hydro technology depends on a watercourse being available on a suitable host site within a community. In the Upper Yarra case study, the host site had been the site of a small hydro plant 100 years earlier that had serviced the early Sanitarium Weet-Bix factory.[31]

Small hydro will require planning approvals. Other legal obstacles, such as biodiversity assessment, tree removal and cultural heritage, should be considered. Earthworks might be required to buffer the sound from neighbouring properties.[32]

The planning of the hydro project took eight years mostly due to regulatory changes that occurred during the project requiring it to be amended.[33] Even though the planning stages and engineering works are more onerous than solar PV, the 100 year lifespan of the infrastructure makes the return on investment attractive for the right location, provided that the significant money needed for upfront investment can be procured.

Community benefits

Despite the relatively moderate profits in small scale projects, many large and sophisticated investors were involved in investing in the projects outlined in the case studies. They were motivated, not only by returns on their investments, but also by the social and environmental gains for the communities. For example, local councils, community funds and local government agencies invested in many of the projects explored in the case studies. In the Hepburn wind project, several million dollars was borrowed from a local bank. In the Upper Yarra Community Energy Project, the community purchased their own bank to fund the project.

Additionally, multiple case studies were researched that attracted large investors that had found novel ways to scale up investment in small infrastructure projects in remote communities. For example, CfAT, by expanding their investments across multiple projects thereby increased their economies of scale and increased their profits. They also attracted more grants the larger that they grew. State and Federal governments are also following this strategy in their latest policy announcements.[34] Other case studies that demonstrated large investors following the strategy of scaling up to multiple projects to increase economies of scale were: Grameen Bank (India), CORENA non-profit revolving funds, Repower (rural Australia) and the World Bank (in the Pacific Renewable Energy Project).

On an even larger scale, publicly listed energy, oil and gas companies and superannuation funds are under increasing pressure from shareholders to invest in renewable energy.[35] This creates an opportunity to attract investment, even in small scale projects, from large investors. This would especially be the case where a series of small-scale projects were linked together as in the Grameen Bank, CfAT and CORENA case studies. Additionally, sustainable bonds, super funds and other kinds of publicly listed green funds are now increasing on the open stock market.[36] These kinds of funds could source sustainable projects for their investment portfolios.

The case studies have demonstrated that both local community investors and large corporate or government investors can realise profits by investing in renewable energy projects in remote communities.

Each remote community project analysed above has provided valuable learnings that inform the overall results. A cost benefit analysis of the case studies demonstrates that investors can be motivated by a range of factors. These include: return on investment and profit, cheaper energy, regional development, local job creation, tourism, shareholder income, education and training outcomes, increased energy efficiency, energy self-sufficiency, development of local assets, resilience, empowerment/self-determination and community outcomes.

Increasingly, investors are motivated by social benefits often demonstrated by an increased standard of living for communities. The provision of fresh water, lighting, safety, powered health centres and school of the air as well as improved hygiene and health with the ability to refrigerate food and medicine. A community infrastructure project must embody a range of investor motivations if the project is to attract investment.

The SDP and Sustainable Community Investment Indicators (SCIIs™) are innovative tools for assessing sustainability of a project both for attracting investment and for monitoring the projects. These are designed out of the benefits that were discovered in the cost benefit analysis of the case studies.

In the case studies, many of the remote community projects were driven by local, small scale community investors. This category of investor is motivated and attracted by social and environmental benefits for the community as well as autonomous and transparent governance. The legal governance structure is also crucial for attracting the type of investors that suit the particular project, whether that be local community investors or large institutional investors. Even where the profits or dividends are moderate, when a project can demonstrate that there is a return on investment and that there is a short payback period on the CAPEX, local community investors are attracted to the project.

Employment was one of the key benefits that both attracted investment and improved standards of living for communities in the case studies. For example, The Desert Peoples Centre near Alice Springs is a joint venture project between the Batchelor Institute of Indigenous Tertiary Education and CfAT.[37] Joint ventures are used to enable different Indigenous Groups to work together on a common project, such as employment and training services. CfAT also has a 50:50 joint venture with MyPathway through a private company. Through this vehicle, they are entering into unincorporated joint ventures with local Aboriginal corporations and are awarded Commonwealth Government funding to deliver projects to Aboriginal Communities.[38] The aim of these joint ventures is to maximise local and Aboriginal employment and education.[39] They are a useful vehicle for engendering cooperation among disparate communities within a region. If managed in a transparent and accountable way, they can be used to attract grants and funding in related industries, such as renewable energy education, training and employment.

Another important community benefit was the increased education and training opportunities that arose. For example, The Australian Capital Territory has a new National Trade Centre at the Canberra Institute of Technology which includes qualifications in renewables skills.[40] If online courses were provided by institutions such as these, it would enable remote training and learning, saving travel costs and increasing the number of people who could complete maintenance training courses.

Some renewable energy providers have an ongoing remote maintenance service. This includes initial training in the maintenance manual and ongoing telephone assistance with following the maintenance manual. Also, there were examples of Australian Indigenous children in remote areas being able to attend school through School of the Air once their communities were electrified. Many energy projects for schools in Africa rely on donations and non-profit organisations for funding. Figure 7.2 shows one such fundraising event.

What follows is a summary list of the benefits appraised in the case studies where sustainable renewable energy infrastructure was built for remote communities. These benefits can be used to demonstrate the attractiveness of a renewable energy investment to financiers.

FIGURE 7.2 A community secondary school in Nigeria, Africa at a donation raising event held by a non-profit organisation. Photo by Emmanuel Ikwuegbu from Unsplash.

Environmental:

- Divesting from coal, gas and coal seam gas. This was an important consideration for farming communities who are worried about the pollution to the water table caused by fracking and gas;
- Communities who want to tackle climate change felt empowered;
- Donations used as offsets for tax purposes or as moral offsets. An example of a moral offset, a local donor will donate to a fund when they take a flight;
- Prevention of deforestation and soil erosion caused by using timber for fuel;
- Using technology that is water efficient;
- Low emissions and low pollution technology used; and
- Integration of solar infrastructure with agriculture such as cattle farms.

Economic:

- Reduced amount of diesel fuel needed for generators, removing the need for expensive transport of fuel to remote communities;
- Increased energy security (less power outages from lack of diesel supply or power card outages);
- Short pay-back periods on CAPEX after 2–3 years;
- Return on investment can be increased if investment in scaled up across multiple projects;
- Population increase, Indigenous people moved from isolated outlying areas into communities with solar power;
- The increase in population caused investment in further infrastructure such as health centres, workshops and school rooms;
- Local businesses flourished;
- Shops are able to open at night; and
- Creation of jobs in electrical maintenance, construction, tourism and community farming.

Technological:

- Hygiene and health has improved through the use of electric appliances such as washing machines and fridges;
- Access to education such as online school/school of the air or renewable energy education centres has increased;
- Reliable affordable electricity; and
- Air-conditioned buildings, in particular, schools were built.

Governance:

- Community involvement in enterprise design was evident;
- Communities benefited from legal governance and funding structures to raise investment money and to sustain ongoing income;

- Communities benefited from legal structures to maintain funds needed for maintenance and upgrades; and
- Legal structures were used to negotiate between communities in relation to employment and training.

The successful renewable energy projects chosen in the case study selection were sustainable across environmental, economic, technological and governance aspects. Therefore, they are sustainable projects or Sustainable Systems.

The sustainability of the system is often evidenced by a standard of living increase and social or community benefits. In the case studies, the following standard of living benefits were identified:

Improved standards of living (community/social benefits):

- Health issues associated with biomass burning such as lung disease and burns from open fires and kerosene lamps have been prevented;
- Access to reliable refrigeration and fresh food;
- Reliable affordable electricity;
- Streetlights are available in remote communities improving safety;
- Households can be productive in the evenings, for example studying;
- Communication has improved through television, mobile phones and radios;
- Access to freshwater is made possible (ability to pump bore water or store and pump rain water or refrigerate hot tap water in desert communities);
- Construction of health clinics in remote communities enabled;
- Creation of jobs in electrical maintenance, construction, tourism and community farming;
- Improved employment and training opportunities; and
- Additional infrastructure and services were attracted to the communities due to the reliable power source.

It is these benefits that were used as a starting point for development the SCIIs™.

Other considerations when attracting investment

Property selection/tenure investigations

One of the main reasons that an infrastructure project fails before it begins is that the host site has not been selected well. The location must suit the technology and consultations with experts, such as real estate agents, valuers, property lawyers, energy and resources consultants, engineers and/or financial advisors are necessary.

When the property assessment is conducted, the expert should uncover any tenure issues. They must discover the background of the owners/landlords/tenants etc. If Indigenous participants are involved, an analysis of their property rights/entitlements in the underlying tenure as well as cultural heritage considerations must be made. The participants in the property negotiations must also be creditworthy. Importantly, the host site must be suitable for the technology chosen. If grid connected, it is necessary to commence negotiations regarding the sale of electricity to the retailer or network provider. The actual construction of the existing infrastructure must also be assessed (roofing structures for solar, proximity of neighbours for wind, suitable river site for hydro, free renewable feedstock for biogas). Any planning issues should be identified upfront by a town planner and property lawyer.

As identified in the case studies, the selection of the host site is crucial to the success of any project.

Due diligence

Due diligence in all aspects of the project is important, for example, due diligence on the installer, the technology used and contract or loan agreement conditions are recommended. A professional lawyer, accountant and electrician will be consulted for most aspects of due diligence. Consumers should also be prudent in conducting due diligence on the installer. Funds should also be set aside for repowering/decommissioning.

The first question to be asked, no matter what technology is chosen, is whether the technology is environmentally sustainable for the location. Non-network (off-grid) renewable energy solutions are best suited to the remote regions of Australia in terms of affordability and efficiency of infrastructure builds. As stated by former Chief Scientist, Dr Finkel, this reduces upfront network costs, increases participation in the market and flattens demand peaks. Non-network solutions also create resilience through diversity as they can operate as standalone or connected systems when designed to do so.[41] The case studies analysed demonstrate that this model is often sustainable; economically, environmentally and technologically. Two case studies that are grid connected are also analysed. These large-scale solar projects in rural Australia have proven to be profitable by selling electricity through PPAs to large buyers such as the Sydney Opera House, Coles and Westpac.

Summary

The development of an infrastructure project can take anywhere from a few months for a small solar PV system, to a few years for a large micro grid.[42]

In the pre-development stage, the host site is assessed. A risk analysis is conducted. Local community sentiment and involvement is explored. At this point, community engagement is essential. Consideration to the diversity within a community and not only the diversity between different communities should be considered. Community engagement recognises that communities are made up of groups of diverse individuals.[43] Planning requirements are assessed. Professional costs including accounting and legal are considered.

The property assessment is also conducted. Are there any tenure issues? Who is the owner/landlord/tenant? Are the participants creditworthy? Is the host site suitable for the technology chosen? If grid connected, have negotiations been commenced for sale of electricity? Has the construction of the existing infrastructure been assessed (roofing structures for solar, proximity of neighbours for wind, suitable water site for hydro, free renewable feedstock for biogas)?[44]

The due diligence must be conducted in relation to the installer, technology and solar contract conditions. A cost-benefit analysis of the size of the system and the Renewable Energy Certificate's (REC's) value is analysed and the customer's capacity to purchase electricity or use electricity on site is been considered.[45]

A crucial step in the pre-development phase is securing investment to fund the project. Once investment is secured the development phase can commence.

In the first part of the development phase, local planning approvals are finalised, host sites are secured, contracts are finalised and equipment is ordered (for example, turbines or panels).

Next, the construction phase will commence. This includes work by project managers, electricians, engineers, construction companies, civil works and labouring.[46]

After the project development is completed, the success of the project is monitored in the post development phase.

At the end of the project lifecycle, the decommissioning phase commences. In this stage, infrastructure is upcycled, refurbished, repowered, upgraded or decommissioned.

The two key stages of a development project that were analysed are:

1. Attracting investment at the pre-development stage; and
2. Monitoring the success of the investment post-development stage.

The next chapter provides the innovative set of indicators, Sustainable Community Investment Indicators (SCIIs™) specifically designed for attracting investment in and monitoring outcomes of infrastructure built in remote and regional communities as opposed to urban areas where indicators have been previously focused.

The SCIIs™ are informed by the benefits identified in the case studies and the SDP which was developed to assist communities in designing a Sustainable System. The objectives of the SDP are to critically analyse investment opportunities in order to test their level of sustainability. Sustainability is defined in the proposition to consist of the following four key objectives:

1. Economically self-sufficient and sustainable, that is they are independently profitable without needing ongoing government funding;
2. Environmentally sustainable;
3. Governed sustainably at a macro and community level; and
4. Technologically sustainable.

When the SDPs are satisfied, there is a Sustainable System or sustainable community, often evidenced by social benefits such as energy reliability and an increased standard of living.

When approaching investors in the pre-development stage of an infrastructure project, the SDP and Sustainable Community Investment Indicators (SCIIs™) can help articulate the benefits of the renewable energy infrastructure. The team of professionals assessing the viability of the development can use the SCIIs™ matrix to flesh out a complete proposal that is attractive for investors. The SDP and SCIIs™ can be used to assess and attract investment from these large investors.

The SDP and SCIIs™ can be used to tailor an investment proposal to different categories of investor, highlighting the strengths and benefits of a community project investment. The SCIIs™ can also be used to monitor the success of an investment in the post-construction phase.

Notes

1 Bulletpoint 2012; SBS news 2020; Australian Government, "Funding to support studies into improved energy supply arrangements"; Vorrath 2020; EWB challenge; Energy Queensland, Annual Report 2017–2018, 51; National Indigenous Australians Agency; Solar Hybrids.
2 CfAT Annual Report, 2017–2018, 13–15.
3 Nehme 2014.
4 Bushlight 2014. Case Study 35.
5 Bushlight 2014. Case Study 31.
6 Ibid.
7 Hicks 2018. *Community Power.*
8 Ibid. 194.
9 Ibid. 179.
10 Coalition for Community Energy 2017. *Small-Scale Community Solar Guide*, v.2.
11 Ibid.
12 Ibid.
13 Desert Knowledge Australia 2019, 24.
14 In Australia, the *Corporations Act* (Cth) 2001.
15 Coalition for Community Energy 2017. *Small-Scale Community Solar Guide*, v.2. 22.
16 Frontier Impact Group 2018, 85.
17 Coalition for Community Energy 2017. *Small-Scale Community Solar Guide*, v.2. 16.
18 Ibid.
19 Indigenous Business Australia, Homepage.
20 Australian Government, National Indigenous Australian Agency.
21 CAT Homelands and Assets Report, 2016, 62.
22 Nehme 2014.
22 Bushlight 2014.
23 Nehme 2014.
23 Bushlight 2014. Case Study 15.
24 Droege 2009, Ch. 6; Australian Government Austrade 2017.
25 Coalition for Community Energy 2018. Webinar 6.
26 CAT Homelands and Assets Report, 2016, 36.
27 Hicks 2018. *Community Power.*
28 Victorian State Government 2015.
29 Coalition for Community Energy 2018. Webinar 7.
30 Ibid.

31 Ibid.
32 Ibid.
33 Ibid.
34 In Australia for example: Bulletpoint 2012; SBS news 2020. Australian Government, "Funding to support studies into improved energy supply arrangements"; Vorrath 2020. EWB challenge; Energy Queensland, Annual Report, 2017–2018, 51; National Indigenous Australians Agency; Solar Hybrids.
35 Klimenko 2019.
36 Potter 2021.
37 Coalition for Community Energy 2018. Webinar 6.
37 CAT Homelands and Assets Report, 2016, 36.
38 CfAT Annual Report, 2017–2018, 14.
39 Ibid. 13–15.
40 Canberra Institute of Technology.
41 Finkel 2017, 72.
42 Hicks and Ison 2018.
43 Eversole 2010, 29–41.
44 Ibid.
45 Ibid.
46 Ibid.

8

SUSTAINABLE COMMUNITY INVESTMENT INDICATORS (SCIIS™)

Introduction

Monitoring a project post completion is an essential part of the planning process. The following section develops a set of original indicators that can be used to either monitor an infrastructure project post-construction phase or that can be used to attract investment in the project pre-construction phase by being used as a decision-making tool. The system of monitoring that has been developed for this purpose is the Sustainable Community Investment Indicators (SCIIs™). The SCIIs™ can be found in Tables 8.1, 8.2 and 8.3. When the SCIIs™ were developed, it became apparent that the Sustainable Development Propositions (SDPs) are fluid and can apply to more than one category of Indicators. Therefore, SCIIs™ are broken into five categories with SDPs applying to each of those categories.

Limitations of similar tools used in other contexts

There are multiple indexes or matrixes used to rank city infrastructure.[1] Any given city can rank very differently depending on the matrix performing the ranking. Singapore, for example, was called the "World's Smartest City" by Forbes Magazine; however, it ranks 52nd in the Economist's liveability rankings, held back by its low ranking in democratic freedom and the environment. Comparatively, Mercer Consulting ranked Singapore as the "Best City for Infrastructure."[2] Therefore, indicators can give varied results depending on the criteria and application. There is no one-size-fits-all set of indicators for smart cities. Despite this, smart city indicators are useful for decision-making in relation to infrastructure investment and are used by investors and governments to decide where there is a gap in infrastructure and where investment should be made to narrow that gap.

DOI: 10.4324/9781003324669-8

FIGURE 8.1 Three light bulbs outlined against a blue sky. Photo by Pixabay from Pexels.

According to the UN Population Fund, city living is still the best indicator of poverty reduction, especially for women, providing higher chances of employment and education and lower rates of child marriage and death in childbirth.[3] Additionally, inner-city apartment blocks are much more efficient than suburban houses. It costs around 80% less to heat and cool. This is because each apartment insulates and heats the apartments on each side, above and below.[4] In addition, the dwellings are less-exposed, so less heat is lost through the windows. The roof has a smaller "heat island effect and resources are pooled and shared (swimming pools, driveways, and stairs)"[5]. In addition, there is less need to use a car in these dwellings if they are located in urban areas close to transport.[6] For these reasons, most of the research and innovation into sustainable living indicators has focused on improving the standard of living in cities.

The other reason that research previously focused on urban growth and city infrastructure is because already over half of the Earth's population lives in cities and this number is growing. Additionally, urban areas consume 75% of worldwide energy production and generate 80% of CO_2 emissions.[7] Much of the growth in the future will take place in communities on the edges of cities as those city footprints expand.

However, it must be remembered that populations in remote and rural communities also increase at a rapid rate when infrastructure is supplied, therefore forward planning when installing infrastructure to account for rural urbanisation should be employed, as has been demonstrated by the case studies. The rural population worldwide is larger than the urban population; however, the trend is for the rural population to be urbanised through migration to cities and urbanisation of rural areas.[8] But, more importantly, and as outlined in the introduction and as recognised in the *Special IPCC Report*, the development and growth in remote indigenous and agricultural communities must be done in a sustainable way as those communities are among the most vulnerable to climate change and lower standards of living.[9]

While smart city indicators have been developed to monitor the impact of smart infrastructure that is built in cities across the European continent, these indicators are designed for application in large cities.[10] Consequently, rural and remote communities have been overlooked creating a gap in infrastructure supply.[11]

Specific indicators designed for sustainable infrastructure development for regional indigenous and rural agricultural communities are needed. Therefore, the following Sustainable Community Investment Indicators (SCIIs™) monitor the impact of sustainable infrastructure development in communities. The proposed innovative system results in a comprehensive and focused use. It can help monitor the outcomes of infrastructure development projects but it can also be used as a starting point for grant applications or discussions between stakeholders and investors or decision-making discussions within communities themselves. Overall, it is a decision-making tool that can be adapted for use in a variety of community contexts.

How a professional uses the SDP and SCIIs™

SCIIs™ are not intended to become policies, laws, regulations or ideologies. They are not a one-size-fits-all procedure or mandate intended to be homogenised globally in the sustainability movement. In this way, they can be distinguished from the ESG (Environmental Sustainable Governance) models that are currently on trend in the European Union. Instead, SCIIs™ are fluid, interchangeable procedures that can be applied depending on the context of the particular project. This is much more relevant for remote communities that are made up of individuals and leaders from a variety of pastoral, indigenous and other backgrounds with unique needs specific to their individual environments and circumstances. One of the primary goals in developing the SCIIs™ was to ensure that the tool could flexibly be applied to all of those different individuals with complementary and competing interests that occupy remote communities often with overlapping tenure interests on the same parcels of land.

Community and industry leaders, with the help of their professional advisors (lawyers, accountants, engineers and valuers), will decide how and when to apply these indicators. Regulation of these indicators is unnecessary as all underlying principles can be found already embodied within western property rights law.[12] Creating regulations for environmental social governance doubles up on the transaction costs for projects cutting small players and large potentially transitional players out of the marketplace. It also reduces investment into smaller projects from large investors as many large institutional investors in Europe will only partner with projects that involve the World Bank as they are guaranteed an ESG rating for their shareholders. Following ESG regulations in Europe, much of the private investment in offshore indigenous communities was withdrawn due to those communities' non-compliance with onerous and discriminatory European ESG regulations.

The SDP has been developed to assist decision-making at the pre-development stage and the Sustainable Community Investment Indicators have been designed to complement this principle and to address the complex task of monitoring the effects of building sustainable infrastructure in remote communities as well as a starting point for discussions with potential investors.

The SDP, when used in conjunction with the SCIIs™, can attract investment in community infrastructure. This can assist rural communities to step into roles of leadership to provide infrastructure for their communities where it is not funded by government due to the vast distances between those communities and the grid system. In brief, this system would help develop communities that are technologically interconnected and sustainable, comfortable and secure.

More traditional methods of monitoring systems are lifecycle analysis and impact assessments conducted by town planners or government officials. These methods have their benefits and are extremely useful within specific contexts. The purpose of the SDP and the Sustainable Community Investment Indicators is to assess, monitor and design complete and complex Sustainable Systems in much broader contexts. Traditional methods focus on one particular aspect, for example, an Environmental Impact Assessment focuses on whether a system is environmentally sustainable. This traditional method does not take into account social aspects, good governance, financial sustainability and technological sustainability. The innovative SDP and SCIIs™ can be used to design and monitor the entire system's sustainability. The SCIIs™ were developed out of the cost-benefit analysis conducted in the case studies and the SDP analysis for this purpose.

The other difference between the SCIIs™ and a traditional method such as an Environmental Impact Assessment is that the SCIIs™ are fluid and adaptable to the communities' individual and unique enterprise design allowing for maximum community involvement. This also means that each community can prioritise which indicators are important to their unique context and enterprise design. As innovation occurs, the indicators remain relevant as no specific examples of "sustainable technology" are listed. Additionally, the SCIIs™ are uniquely designed to attract investment.

The SCIIs™ are summary tables of indicators for monitoring the success of a community renewable energy infrastructure project. The rating under the SCIIs™ will depend on the different weights applied to the criteria and the information and data input into the model. This in turn depends on the quality of professionals employed to undertake this task.

Therefore, it is important to consider, which experts will oversee the assessment, what financial resources are available to undertake the assessment, who are the appropriate decision-makers, what specialists and experts are required, what community stakeholders should be participating, what is the timing of the assessment, what is the goal or purpose (attracting investment, monitoring post development, or at the planning stage), what data and information is available and how will the assessment process be evaluated.

When involving community stakeholders, several methods can be used. Some examples include surveys, polling, town meetings, festivals, conferences, interviews, expert panels, focus groups, planning groups and interactive apps.[13]

When used effectively, SCIIs™ can be used as a starting point for grant applications, investor negotiations, discussions between stakeholders or decision-making discussions within communities themselves. The primary advantage of using this tool is that complex decisions can be made. It is a decision-making tool that depends on the people who are using it and the data that they input into the tool.

The above discussion on sustainability pertains to these key areas:

1. Environmental sustainability, including sustainable yields of natural resources;
2. Sustainable governance;
3. Economic sustainability; and
4. Sustainable technology.

By following this process, a Sustainable System that provides community benefits and an increased standard of living can be developed.

The analysis of these areas can be summarised by a statement of propositional logic or propositional calculus. Propositional logic is the bridge between conceptual thought, language and mathematics. It is used for purposes as varied as computer programming and drafting legislation. The SDP proposition has been formulated for the purpose of enabling a team of interdisciplinary professionals (or one interdisciplinary professional with sufficient external advisors) to make decisions on infrastructure projects. The success with which it is used depends on the quality of information and expertise that each advisor brings to the decision-making process.

Propositional logic algorithms are used to create systems that can be applied to a variety of facts and circumstances with consistent results. The application of the algorithm is dependent on the facts and circumstances of the scenario. For example, the social conditions, facts or circumstances between two different case studies could vary and therefore the results obtained when applying the algorithm would vary between case studies. The SCIIs™ are a matrix of indicators designed to assist expert advisors with the input of facts and circumstances for each individual case that is assessed. This ensures that each relevant fact is considered if and when the conditions of each infrastructure project vary, thereby giving accurate results as to the sustainability of investment in a project.

Risks

These indicators are not intended as an oracle. The success of their deployment depends on the quality of data being fed into the model by expert professionals in the various disciplines as need be.

> Before going into more detail, it is appropriate to mention the observer effect. There is an "observer effect" in many fields of research where a researcher influences the results by their mere presence in the experiment.[14] It is important to be conscious of the scientific fact that observing can cause change in the system and so monitoring must be done only in a limited way

where necessary and within the rights of the population monitored such as privacy and consumer data right legislation.

On the other hand, monitoring is necessary to amend errors in design, "make the wrong choice in the design of our smart cities, and our descendants may find themselves a century out, wondering what we were thinking today."[15] For example, concentrations of air pollution such as lead are found in Greenland ice showing historical spikes of pollution during the industrialisation of Rome. Metal mining in Europe extends back at least 5000 years in Spain and Portugal. Iberian copper, lead, iron and silver deposits were worked by Tartassians, Phoenicians, Carthaginians and Romans:

> The residue of millennia of strip mining and smelting has left its mark in the soil of southern Iberia. Indeed, it has left its mark on the world. In 1997, a team of environmental scientists found traces of Iberian atmospheric lead pollution in the Greenland ice core .... The airborne lead pollution around the mines in fact left a chemical residue in the bones of the Iberian people.[16]

Historical analysis of pollution can inform current decisions on infrastructure development and ensure future generations have a healthy environment in which to build communities.

Additionally, being counted is a fundamental act of inclusion. Lacking basic information about the people living in a community makes it difficult for the community to assert its rights and its needs. For needs to be supplied, those needs must be measured and mapped.[17] For example, an open digital street map project in Kenya was used to focus on powering change in the community. It recruits residents to report on the progress of infrastructure projects. This has led to the installation of water pumps and latrines where sewerage is unavailable and allows the government to effectively monitor and audit contractors. Residents can report on works in progress. In this example, the community is viewed as being made up of real people rather than numbers or statistics.[18]

A liveable community is a fine balance between community spirit and anonymity. People want to live near neighbours that are neighbourly but they don't want to feel coerced, interfered with, gossiped about or gazed upon. They want leafy streets where they can sit safely and chat to each other and to live in places where they can go about their personal lives. They want schools, offices, shops and restaurants with pedestrian access and affordable transport. Affordable energy, healthcare services and public transport are important. Participation in decision-making and governance is also needed.[19]

How the SCIIs™ were developed

The idea for the SCIIs™ was born from the lack of Australian community-specific indicators as opposed to the plethora of smart city indicators in Europe. The starting point for the development of the SCIIs™ was the cost-benefit analysis conducted in the case studies. The SCIIs™ encompass all of the primary benefits of community renewable energy projects and the potential pit falls or costs evaluated in the case studies.

The framework for the SCIIs™ was based on the literature review of the SDP and the case studies. The literature review identified the four main objectives of a sustainable community project, namely, sustainable governance (including macro governance, legal structure sustainability and community governance), economic sustainability, sustainable technology and environmental sustainability. The complexity of meeting the definition of sustainable in each of these categories is apparent and should not be abbreviated or codified. Due to its complexity, it will be the topic of debate among differing community participants and leaders. Where all of the SDPs are satisfied, a Sustainable System or community exists and enjoys all of the social benefits that come along with an increased standard of living, such as improved energy access, health, education and employment.

The set of Sustainable Community Investment Indicators was developed to ensure that an investment decision was thorough and unbiased in relation to each of these categories of sustainability. For example, under the category "economic sustainability," it was found in the case studies and literature review that not only is

FIGURE 8.2 Latin American solar technician installing roof panels. Photo by Los Meuertos Crew from Pexels.

micro-economic sustainability required to attract investors (the return on investment and CAPEX payback periods) but macroeconomic sustainability must also be considered. It was discovered that the natural capital aggregate rule must be upheld in a sustainable community so that the renewable and non-renewable resources available to a community are not depleted beyond sustainable yields for future generations living in that community. It was findings such as these that informed the development of the SDP and comprehensive matrix of SCIIs™ designed to collect and monitor the data fed into the decision-making tool.

Sustainable Community Investment Indicators

The decision tool below is an example of soft technology[20] and works best in a free-market private ownership framework or community ownership project. The investment decisions of each community, as shown in the case studies, will be context specific and the decisions made must be free choices to enable self-determination and successful design participation by community members in a free-market democratic environment.

Table 8.1 is an example of a rating system that can be used to analyse the success of the investment motivators in Table 8.3. Users can rate from 0 to 5 sustainable governance, sustainable technology, sustainable economy, sustainable environment and sustainable community.

Table 8.2 provides descriptions that can be applied to the rating system at Table 8.1 above. For example, a rating of 0 means sustainability is not satisfied, whereas a rating of 5 means that sustainability is satisfied long term across all of the SCIIs™ in Table 8.3 below.

*Sustainability is defined by the SDP.

$$((NCNC \vee NC) \wedge IG \wedge PP \wedge DRL \wedge (POP \leftrightarrow EG) \wedge HR \wedge T \wedge (POP \leftrightarrow SL)) \rightarrow SS$$

TABLE 8.1 SCIIs™' Status Table

Status	*SCIIS™*				
	Sustainable Governance	*Sustainable Technology*	*Sustainable Economy*	*Sustainable Environment*	*Sustainable Community*
5					
4					
3					
2					
1					
0					

Table 8.3 contains examples of data that can be used to inform the SCIIs™ Tables 8.1 and 8.2 above. Not all of the indictors in Table 8.3 will be present in a singular project. The usefulness of the indicators is to provide motivation for best practice project design, decision-making and investment.

Table 8.3 contains Sustainable Community Investment Indicators that can be used in development projects. They will be applied in hypothetical case studies in the next chapter.

TABLE 8.2 SCIIs™' Status Description Summaries

Status		*Description*
5	Long-term sustainability	Sustainability is satisfied long term across all SCIIs™.
4	Medium-term sustainability	Sustainability is satisfied medium-long term across the majority of SCIIs™.
3	Short-term sustainability	Sustainability is satisfied in the short term across some SCIIs™.
2	Compromised sustainability	Sustainability is compromised across the SCIIs™.
1	Limited sustainability	Sustainability is not satisfied in the majority of SCIIs™.
0	No sustainability	Sustainability is not satisfied.
1–2	Actions	A rating of less than 4 requires improvement to and adjustment of the system to reach the goal of a Sustainable System.

TABLE 8.3 SCIIs™ Investment Motivators/Monitoring Indicators

SCIIs™	*Example Benefits*
Sustainable governance	General governance principles:
The democratic rule of law (DRL) is upheld **Λ DRL (fourth proposition)** **Human rights and liberties (HR) are upheld** **Λ HR (fifth proposition)**	• Independent, accountable and transparent decision-making; • Local participation/engagement in decision-making; • Democratic governance; • Public and social services provided; and • Public health and safety standards. Democratic legal governance: • Separation of powers doctrine upheld; • Democratic rule of law upheld; • Access to legal justice system; • Human rights upheld; • Protection of the right to liberty; • Equality before the law;

TABLE 8.3 Cont.

SCIIs™	*Example Benefits*

	• Legal certainty;
	• Processes for resolving disputes;
	• Prevention of discrimination;
	• Civil and political rights upheld;
	• Economic, social and cultural rights respected;
	• Anti-corruption;
	• Fair competition;
	• Access to justice;
	• Respect for property rights, both tangible and intangible, in particular any compulsory public acquisitions are done on just terms for fair compensation;
	• Fair marketing, factual and unbiased information;
	• Fair contractual practices;
	• Protecting consumers' health and safety;
	• Consumer service, support, and complaint and dispute resolution;
	• Warranties enforceable; and
	• Access to justice and essential services.
	Accountable and sustainable organisational governance:
	  • Due diligence conducted.
	  • Contractual, corporate, financial, organisational and legal structures are sustainable.
	  • Communities and businesses have freedom of choice over enterprise design.
	• Safety and security including any impact on country risk is considered; and
	• Cooperation and trade with other communities/societies/countries.

Sustainable economy
Economic growth (EG) increases in line with population growth (POP)
Λ (POP ↔ EG) (sixth proposition)
The standard of living increases in line with population growth (POP)
Λ (POP ↔ SL) (eighth proposition)
	• Funding arrangements and/loans are serviceable.
	• Innovation:
	  • Improved level of technological innovation and entrepreneurship.
	• Employment:
	  • Local job creation;
	  • Sustainable employment and flexibility of labour market;
	  • Employment conditions and social services;
	  • Health and safety at work;
	  • Human development and training in the workplace; and
	  • Productivity improvement.
	• Improved capability to trade.
	• Considered operational costs and capital expenditure in the ROI analysis.

(continued)

TABLE 8.3 Cont.

SCIIs™	*Example Benefits*

	• Investment attractiveness (ROI and payback periods can be cross-checked with other valuation methodologies).
	• Transaction costs are serviceable.
	• Tax burden vs revenue source is sustainable.
	• Maintenance of productive capital.
	• Education and training opportunities.
	• Competitiveness in the market and consumer protection.
	• Limit public debt: Either at the expense of future generations or at the expense of the independence of Parliament.
	• Wealth and income creation:
	  • Local industry (for example, renewable energy maintenance and construction, manufacturing or tourism);
	  • Income to shareholders/ community;
	  • Local jobs/contracts;
	  • Income creation; and
	  • Development of local infrastructure assets.
	• Economic:
	  • Micro and/or circular economies created locally within the community or society.
	  • Funds are able to be raised or set aside for maintenance, upgrades or decommissioning.

Sustainable **Environment** **Natural capital (NC) is enhanced, substituted, adapted OR remains constant** **((NCNC V NC)) (first proposition)** **Intergenerational equity (IG) remains constant** **Λ IG (second proposition)** **The precautionary principle (PP) is satisfied** **Λ PP (third proposition)**	• Application of the environmental aspects of the Sustainable Development Proposition, in particular maintenance, substitution, adaptation or improvement of the value of natural capital. That is, natural capital aggregate rule is met. • Sustainable renewable energy sources maintained. • Sustainable clean water source and clean groundwater/ water tables. • Prevention of pollution (water, air). For instance, prohibit acidifying, petrochemical pollutants, harmful air pollutants that may affect human health, damage to crops, pollutants such as lead, erosion or deterioration of the soil. • Reduction in greenhouse gas emissions; or • Ensure limited increase in greenhouse gases (e.g., carbon dioxide, methane, hydrogen) in the atmosphere. • Climate change mitigation and/or adaptation. • Protection of biodiversity. • Restoration and maintenance of natural habitats.

TABLE 8.3 Cont.

SCIIs™	*Example Benefits*

Sustainable technology
Technology (T) improves
Λ T (seventh proposition)
Natural capital (NC) is enhanced, substituted, adapted OR remains constant
((NCNC V NC) (first proposition)
The precautionary principle (PP) is satisfied
Λ PP (third proposition)
	• Financial provision and technological consideration have been made for maintenance, upgrades and/decommissioning.
	• Sustainable renewable energy source:
	  • Energy self-sufficiency;
	  • Energy security (reliable and affordable); and
	  • Energy efficiency.
	• Sustainable regulation.
	• Technology development and access.
	• Waste technology is recycled where possible.

Sustainable System or Sustainable Community **(propositions 1–2)**	• Social benefits from an increased standard of living: • The standard of living, income and employment increases over time as the population increases; • Infrastructure availability improves (e.g., removal of energy poverty and increased energy affordability and reliability); • Health conditions; • Solidarity between generations; • Social and cultural facilities; • Community participation; • Education; • Employment creation and skills development; • Gender and racial equality; • Empowerment; • Preservation of cultural heritage; • Both the community and the investors benefit from the development project; • Cooperation and trade with other communities/societies/countries; and • Freedom of choice in enterprise design for investors and communities.

Summary

In order to improve the success of a project, it must be monitored and audited. This is also necessary for the purpose of maintenance, upgrades and refurbishments, decommissioning, repowering or upcycling the infrastructure at the end of the life cycle.

If the infrastructure is to be managed effectively, its functioning must be monitored. It is also a useful tool to attract initial or subsequent investment in

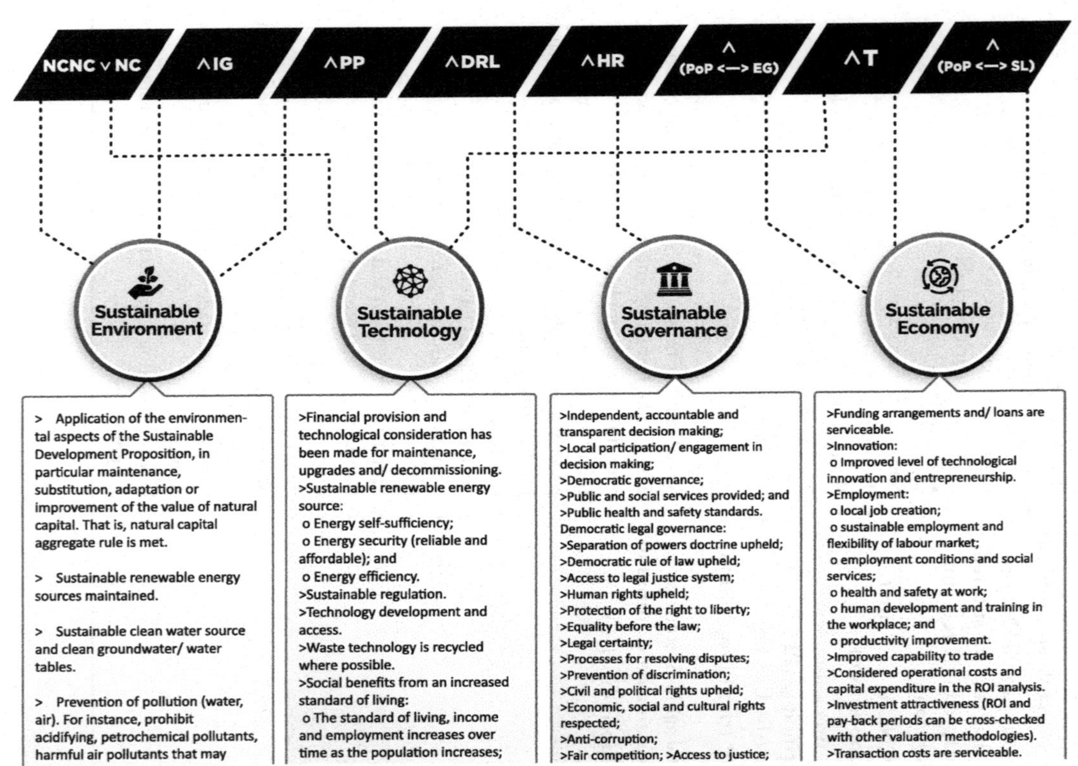
A Sustainable System (SS)
NCNC ∨ NC
∧IG
∧PP
∧DRL
∧HR
∧ (PoP <—> EG)
∧T
∧ (PoP <—> SL)
Sustainable Environment
Sustainable Technology
Sustainable Governance
Sustainable Economy
> Application of the environmental aspects of the Sustainable Development Proposition, in particular maintenance, substitution, adaptation or improvement of the value of natural capital. That is, natural capital aggregate rule is met.
> Sustainable renewable energy sources maintained.
> Sustainable clean water source and clean groundwater/ water tables.
> Prevention of pollution (water, air). For instance, prohibit acidifying, petrochemical pollutants, harmful air pollutants that may
>Financial provision and technological consideration has been made for maintenance, upgrades and/ decommissioning.
>Sustainable renewable energy source:
o Energy self-sufficiency;
o Energy security (reliable and affordable); and
o Energy efficiency.
>Sustainable regulation.
>Technology development and access.
>Waste technology is recycled where possible.
>Social benefits from an increased standard of living:
o The standard of living, income and employment increases over time as the population increases;
>Independent, accountable and transparent decision making;
>Local participation/ engagement in decision making;
>Democratic governance;
>Public and social services provided; and
>Public health and safety standards.
Democratic legal governance:
>Separation of powers doctrine upheld;
>Democratic rule of law upheld;
>Access to legal justice system;
>Human rights upheld;
>Protection of the right to liberty;
>Equality before the law;
>Legal certainty;
>Processes for resolving disputes;
>Prevention of discrimination;
>Civil and political rights upheld;
>Economic, social and cultural rights respected;
>Anti-corruption;
>Fair competition; >Access to justice;
>Funding arrangements and/ loans are serviceable.
>Innovation:
o Improved level of technological innovation and entrepreneurship.
>Employment:
o local job creation;
o sustainable employment and flexibility of labour market;
o employment conditions and social services;
o health and safety at work;
o human development and training in the workplace; and
o productivity improvement.
>Improved capability to trade
>Considered operational costs and capital expenditure in the ROI analysis.
>Investment attractiveness (ROI and pay-back periods can be cross-checked with other valuation methodologies).
>Transaction costs are serviceable.

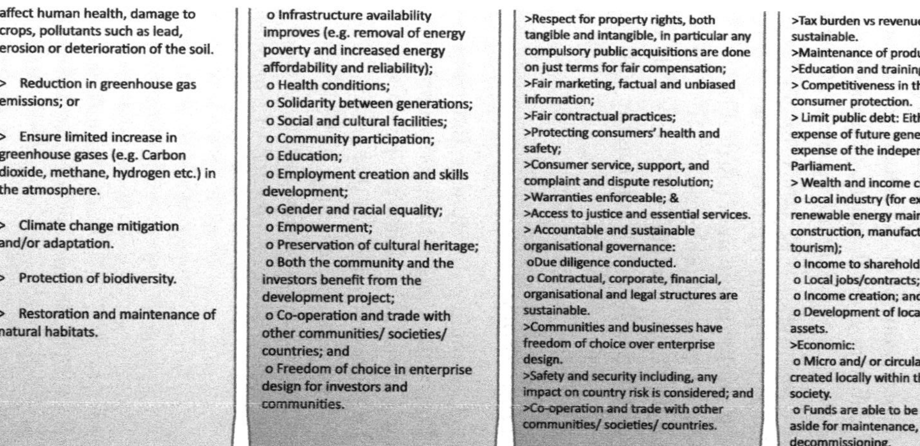

FIGURE 8.3 Diagram showing how the Sustainable Community Investment Indicators in Table 8.3 relate to the Sustainable Development Proposition.

a project as it can explain the potential success of a project, its strengths, support levels, investment models and what planning has been achieved thus far to investors.

A community infrastructure project is made up of local people. They are not only the foundation but also a resource for investment both in money and in time especially as volunteers (this is often referred to as sweat equity). Engaging with the local community can deliver ongoing investment in the project. Community members usually welcome an opportunity to invest in their own community.

As demonstrated by the case studies, investors can be motivated by a range of factors. For instance, profit, cheaper energy, energy security, regional development, local job creation, tourism, shareholder income, education and training outcomes, increased energy efficiency, energy self-sufficiency, development of local assets, resilience, empowerment/self-determination and community benefits. SCIIs™ must embody those investor motivations if the project is to attract investment in sustainable infrastructure development for communities. Projects should also be familiar with the SDP if they are to fall within the definition of sustainable.

SCIIs™ can include data and information collated using the SCIIs™' tables. Communities can also develop their own data collection tables. Figure 8.3 shows how a Sustainable System as set out in the SDP (see Table 1.1) informs the SCIIs™ as shown in Table 8.3.

Sustainable Community Investment Indicators

Figure 8.3 shows how the Sustainable Community Investment Indicators in Table 8.3 relate to the SDP. Each of the Propositions interrelates, for example, sustainable technology must be economically sustainable so that it can be maintained or upgraded. Each of the Propositions also relates to Indicators that are outcomes demonstrating the sustainability levels of the system. For example, if the development project reduces pollution and greenhouse gases as well as maintains natural capital these indicators would demonstrate a level of environmental sustainability, satisfying part of the SDP.

Notes

1 See, for example, Lazaroiu and Roscia 2012.
2 Hollis 2013, 41 from Clark, G. and Moonen, T., "The Business of Cities: City Indexes in 2011," www.businessofcities.com. 12-3
3 Hollis 2013, 337 from UNFPA Youth Supplement, 2007, 7.
4 Ibid. 302.
5 Ibid. 302.
6 Ibid. 303.
7 Lazaroiu and Roscia 2012.
8 Townsend 2013, 1.
9 The United Nations, *Special Report*. 2018, 40.
10 European Commission, "Standards: City – Smart cities".

11 Hicks and Ison 2018 with reference to; Harnmeijer et al., 2012; Hicks and Ison, 2011; Kildegaard, 2010; Kildegaard and Myers-Kuykindall, 2006; Kirsch et al., 2015; Lantz and Tegen, 2009; Middlemiss and Parrish, 2010; Mulugetta et al., 2010; Seyfang et al., 2013; Walker et al., 2007a; Walker, 2011.

12 See, for example; *Environmental Protection Act* (1994) (Qld); The *Planning Act* (Qld) (2016), s 5; *Universal Declaration of Human Rights; The Commonwealth Constitution of Australia; The Australian Consumer Law.*

13 Townsend 2013.

14 Note: when monitoring community projects, beware of the scientific principle; the observer effect which states that the observer changes that which is observed even without the observed knowing that they are observed. An experiment conducted in quantum physics proved that the presence of an observer, even if that observer was an electronic instrument impacting on the results of the experiment, even at the quantum level. From Weizmann Institute of Science (February 27, 1998). "Quantum Theory Demonstrated: Observation Affects Reality." Science Daily.

15 Townsend 2013, 107.

16 Goetzmann 2017, 133.

17 Ibid.185.

18 Ibid.187–188.

19 Girardet 2008, 167.

20 Hallett 2011, 99.

9

HYPOTHETICAL CASE STUDIES

The following hypothetical case studies are designed to demonstrate the application of the Sustainable Development Proposition (SDP) and Sustainable Community Investment Indicators (SCIIs™). They are designed for pedagogical training and learning purposes. For a summary of the concepts used in these exercises refer to Tables 1.1, 8.1–8.3 and Figures 1.2, 5.5 and 8.3.

Case study 1

1. Read the case study "Green Ville" below.
2. Analyse the Sustainability of Green Ville using SCIIs™.
3. Draw a conclusion as to the sustainability of the community as per the Sustainable Development Proposition.
4. Assess whether Green Ville is a viable candidate to connect to the Federal Super Grid.

Case study: Green Ville

---

1. Jurisdiction:

Green Ville is a rural community located beyond the Black stump in regional Australia.

2. Background of community:

Green Ville has an enthusiastic community keen to embrace renewable energy technology and become 100% carbon neutral. They want to attract jobs and industry to their town. The town is remote and has no grid access.

DOI: 10.4324/9781003324669-9

3. Technology used:

The community decides to undertake a SCIIs™ assessment based on the introduction of:

a. Solar PV for all municipal buildings.
b. Biogas for waste treatment and landfill sites as well as the local cattle stations.
c. Wind energy installed on the sheep farms located on the fringes of town and in the hilly, high wind area.
d. Solar PV for households and small businesses.

4. Governance and funding structures:

The community sets up a private company called, ReNewIt Pty Ltd. They decide to raise capital by issuing shares and applying for grants. They intend to set up a community trust fund. A proportion of all income from ReNewIt goes into the trust fund for maintenance and upgrades of the systems. ReNewIt pays annual dividends to its investors out of the profits.

ReNewIt secures a PPA with the local council for provision of solar PV. The ROI model attracts 500 local community members who all show interest in investing. $200,000 capital is initially needed but only 20 investors are allowed under the *Corporations Act* 2001. Therefore, 20 investors are needed with $10,000 capital to invest each. The investment capital is raised from community investors. RenNewIt uses the capital to build a solar farm to supply the municipal buildings in town under the PPA. The remaining interested community members seek out other investment opportunities as follows:

- Micro Pty Ltd raises capital to build micro virtual grid infrastructure.
- Gasso Pty Ltd raises capital to build a biogas plant on the local council land fill site. They secure a lease with the council. They supply electricity to the new virtual grid.
- The Jones family plans a biogas plant on their pig farm and negotiates a private FiT agreement with Micro Pty Ltd to supply electricity to the virtual grid. They create a family trust to govern the business.
- A group of community investors set up a cooperative solar farm and lobby the Federal Government to change the rules on cooperative energy. Rule changes are made to allow cooperative members to benefit from reduced energy prices.

The Jones family uses the extra income to buy a helicopter to run tourist flights.

Their eldest son, Mick, one of seven brothers starts making biofuel from leftover chip oil at the local restaurant, Chippies. He meets the local waitress Jill, who is one of seven sisters. They get married.

The Federal government announces a tax exemption on electric vehicles, 80% of the community buy EVs. Mick sells biofuel to the remaining 20% of community members to run their diesel vehicles. Mick sells the family helicopter and upgrades to an electric powered personal plane. Tourism triples, everyone wants to try out the zero-emissions electric plane.

Jill buys an EV road train that has a graphene sheet battery bank for electrical storage. She uses the road train to create a mobile grid and transports excess renewable energy to neighbouring PittsVille, a community named after the large coal pit outside the town.

PittsVille is embroiled in a court battle with the former directors of Oke Ltd, a coal company that is under administration. Oke Ltd is in the midst of cleaning the pit water in its coal mines and decommissioning its coal fired power plant. Oke Ltd goes into voluntary administration. Many of its former employees seek jobs in neighbouring Green Ville.

Many Green Ville community members are now receiving an income from their investment in the private renewable energy companies. Other community members have taken online TAFE courses to gain skills in electrical trades, maintenance and installation of renewable energy.

The local council partners with the State government to fund an education centre for electrical trades and renewable energy trade certificates. The community upskills. More people from PittsVille migrate to Green Ville for the employment and education opportunities.

The property values increase by 35%.[1] The tax base increases. Property owners borrow against their equity and invest in property improvements, including energy efficiency measures. The local council builds a cultural heritage museum and a health centre with the increase in property rates revenues.

Jill and Mick enrol in a family planning session at the new health centre. They give birth to approximately 1.73 children.[2] Despite the decline in their family's fertility rate, the economic growth (EG) of the community and the resultant migration from PittsVille results in an overall positive population growth rate for Green Ville.

The State Government invests in a solar panel and battery recycling plant in the town, creating further employment opportunities. They enter into a Memorandum of Understanding (MOU) with the Federal government to secure ARENA funding.

The community wants to promote itself as a sustainable town in order to attract further investment. The community members want to undertake a SCII assessment.

You are engaged, advise the community members.

Example answer 1

Below is an example solution to the above exercise.

Sustainable governance

Green Ville is an Australian town.

Australia is governed by democratic law and thus the macro governance is assumed to be sustainable on the facts given.

Organisational governance

On a micro scale, the community has created several private companies to govern their infrastructure project:

- ReNewIt Pty Ltd supplies solar power to the municipal buildings under a power purchase agreement.
- Micro Pty Ltd builds a micro virtual grid.
- Gasso Pty Ltd builds a biogas plant on the council owned landfill site. They lease the site from the Council. They sell the electricity back to the grid. They have agreements with Micro Pty Ltd (the virtual grid network provider).
- The Jones Family Trust builds a biogas plant on their pig farm. They negotiate a private FiT agreement with Micro Pty Ltd to supply electricity to the virtual grid. They also start a tourist business selling joy flights.
- Mick Jones makes and sells biofuel as a sole trader.
- Jill Jones sells electricity using a mobile grid, as a sole trader.

Private companies

Private companies are a simple governance structure that allows capital raising among 20 investors. Although the structure limits the number of investors, it also limits the upfront and ongoing costs of running the structure. The reporting requirements, for example, are less onerous than a public company. Overall, this structure allows community members to join together to raise capital to build local renewable energy infrastructure.

Family Trust

The Jones Family Trust allows the family to use its farm to build a biogas plant and it is assumed on the facts that this allows any income to be distributed to the beneficiaries, namely, the members of the family living on the farm, in a fair and equitable manner. This income also allows them to expand into a tourism business or to branch off as sole traders.

Sole traders

The sole traders have less access to capital raising however they have the flexibility to start their business and the autonomy to run the business themselves and retain all of the income from the business without having to distribute it to other investors. Depending on the nature of their business, there could be more risk in being a sole trader.

General governance

In general, many examples of good governance can be found on the facts:

- Competitiveness in the market was created as multiple renewable energy companies flourished.
- It is assumed on the facts that the companies employed professional advisors who undertook due diligence and ensured accountability.
- It is assumed on the facts that the local council was independent, accountable and transparent in their decision-making.
- The local council provided health and education services as the town flourished.
- The locals participated in the decision-making.
- The Australian doctrine of the separation of powers continued to deliver on all of the principles of democratic legal governance and international human rights protections.

In summary, it is likely that the democratic rule of law (DRL) was upheld and that human rights and liberties are upheld. Additionally, the governance of the micro systems will lead to a sustainable economy as discussed below. Therefore, an analysis of the Sustainable Community Investment Indicators leads to the tentative conclusion that the SDP is upheld as follows:

The democratic rule of law (DRL) is upheld
Λ DRL

and

Human rights and liberties (HR) are upheld
Λ HR.

Sustainable economy

All of the governance structures allowed for income producing enterprises. The private companies are able to distribute income as dividends to the shareholders/investors. Other income producing structures included a private FiT agreement made with the virtual grid company and power purchase agreements that bring in income for investors.

ReNewIt Pty Ltd is arguably the most sustainable funding structure as they created a trust in which to place ongoing portions of the profits for future maintenance, upgrades or decommissioning of the structure.

All of the structures enabled local jobs, an increase in training, productivity and local industry. Trade with neighbouring towns increased. A workforce from neighbouring towns was attracted. Property prices increased and the tax base therefore increased.

No public debt was created. This is important for economic sustainability between generations, also known as intergenerational equity (IG).

In summary, the locals gained income and wealth creation through:

- Local industry (for example, renewable energy maintenance and construction, manufacturing or tourism);
- Income to shareholders/community;
- Local jobs/contracts;
- Shareholder income; and
- Development of local infrastructure assets.

Mick's biofuel industry also created a circular economy.

In summary, it is likely that EG will increase in this community as the population growth increases. The fertility rate is negative but migration is driven by local job creation. Therefore, as jobs are created, a workforce is attracted and EG increases with the population increase.

Therefore, an analysis of the Sustainable Community Investment Indicators leads to the tentative conclusion that the SDP is upheld as follows:

Economic growth (EG) increases in line with population growth (POP) Λ (POP ↔ EG)

Sustainable environment

All of the industries established are based on renewable energy sources. This means that they are using renewable natural capital (NC) resources such as solar, wind and biogas. The pollution created during the energy capture process is negligible if materials are recycled.

The biogas is a by-product of decomposing landfill. It is sustainable because it is not created out of unsustainable biomass.

The biofuel creates a circular system as it recycles old chip oil into biodiesel for vehicles. The chip oil would otherwise be discarded and wasted in a linear system, causing pollution.

Renewable energy sources enable:

- Energy self-sufficiency (they cannot be depleted); and
- Energy security (reliable and affordable).

Renewable energy sources can also assist renewable natural capital such as natural forests, clean water, clean air and local ecosystems in being sustained.

Conversely, in the neighbouring Pitts Ville, the renewable NC (water table, forests and air) was depleted beyond sustainable yields due to the pollution caused by coal mining. The value of the NC was therefore depleted.

In Green Ville, the NC resources were maintained at sustainable yields and the value of NC was therefore maintained for future generations.

In summary, an analysis of Sustainable Community Investment Indicators leads to the tentative conclusion that the SDP is upheld as follows:

The renewable energy industry means that no NC reduction occurs in the energy production for the town satisfying:

Natural capital (NC) is enhanced, substituted, adapted OR remains constant
((NCNC V NC)

The water table, air and forests remain unpolluted due to the lack of coal and gas industry in the town.

Natural capital (NC) is enhanced, substituted, adapted OR remains constant
((NCNC V NC)

The renewable and non-renewable resources are passed on to the next generation intact:

Intergenerational equity (IG) remains constant
Λ IG

On the facts, no environmental risks are taken.

The precautionary principle (PP) is satisfied
Λ PP

Sustainable technology

From a financial perspective, we have already seen that sustainable finance has been set up for one of the companies for maintenance and repowering. This means that the technology is sustainable from a financial perspective because it can be maintained, upgraded or decommissioned where necessary out of those funds. It is not clear on the facts whether all of the companies have put a similar system in place.

On the facts, the technology is assumed to be sustainable as follows:

- Housing was improved for energy efficiency.
- EVs and biofuels were used for transport.
- Reliable electricity enabled reliable ICT.
- Recycling of methane to biogas and chip oil to biofuel occurred.
- All aspects of privacy law, competition law and fair contract law were upheld.
- A virtual grid was built.

In summary, an analysis of Sustainable Community Investment Indicators leads to the tentative conclusion that the SDP is upheld as follows:

Technological advances occurred (EVs, electric road trains, electric planes, virtual grids).

Technology (T) improves
Λ T

The technology does not pose a risk to the health of the people or the environment.

The precautionary principle (PP) is satisfied
Λ PP

The energy used is renewable (wind, solar, biogas).

Natural capital (NC) is enhanced, substituted, adapted OR remains constant
(NCNC V NC)

Sustainable community

An improvement in all aspects of social and health indicators has been demonstrated in Green Ville. For instance:

- Infrastructure was built (health and education centres, virtual grid).
- The standard of living, income and employment increased over time as the population increased;
- Solidarity between generations improved with the principle of Intergeneration Equity being upheld;
- Education was made more readily available by the new education centre;
- A cultural centre was built;
- Employment creation and skills development occurred;
- Empowerment of the community members occurred through wealth creation and improvement in employment and education opportunities;
- Both the community and the investors benefited from the development projects.
- Cooperation and trade with PittsVille occurred.

The population increased due to migration from PittsVille. The standard of living increased as wealth was created, housing was upgraded and increased in value. Infrastructure was built. Health, education and economic opportunity improved across generations.

In summary, an analysis of Sustainable Community Investment Indicators leads to the tentative conclusion that the SDP is upheld as follows:

Intergenerational equity (IG) remains constant
Λ IG
The standard of living increases in line with population growth (POP)
Λ (POP ↔ SL)
Tentative conclusion:

Green Ville is a Sustainable Community according to the SDP.

((NCNC V NC) Λ IG Λ PP Λ DRL Λ (POP ↔ EG) Λ HR Λ T Λ (POP ↔ SL)) → SS

The Sustainability Community Investment Indicators listed above demonstrate the town's investment attractiveness and stability as a sustainable community. Green Ville is likely to attract further investment from likeminded investors and government agencies.

Example 2 (short answer question)

Desert Home Station (hypothetical indigenous community scenario)

The Native Title Registered Corporation, X Group, consists of three extended families living at Desert Home Station. X Group's infrastructure includes:

- Four houses built in the 1970s;
- A manual water bore;
- An outhouse; and
- A diesel generator.

X Group's diesel generator is old. The community would like to investigate community owned solar as a replacement for the diesel generator.

Nearby, Y Group has a solar powered community and has offered to supply solar infrastructure to X Group. Y Group offered to lease panels to X Group for an ongoing fee. They also offer ongoing maintenance for the batteries and panels. The solar infrastructure would transfer in ownership to X Group at the end of the lease term.

There is a history of conflict between X Group and Y Group and X Group would prefer to own their own panels rather than lease them but they do not have the funds to supply their own infrastructure and they do not have anyone in their community trained to maintain solar panels. Y Group has several people trained in solar maintenance.

You are the advisor to X Group. Provide a 2–2 paragraph answer to X Group's request for advice.

Example answer 2

Initial grants

X Group has a variety of options available to them. X Group could apply for a grant through the federally funded agencies such as ARENA, IBA, ABA or Outback Power Program.[3] Or they could apply for grants through Aboriginal Land Councils or State and Territory government agencies. Grants can also be provided by local or international NGOs and universities.

In this situation, the housing could benefit from refurbishment or rebuild. Initial grant applications should be made to Australian government's national partnership on remote Indigenous housing or the Northern Territory's initiative, Working Future. The Federal Department of Families, Housing, Community Services and Indigenous Affairs is also proactive in community grant funding for new housing. Most State governments also have funding available for community infrastructure.

It is important to consider the benefits of partnering with public bodies for grants as they can also assist with building new detached zero emissions housing for communities. There are now detached dwelling houses available that combine design, insulation, energy -saving appliances and PV solar panels to create zero net energy housing. Often the detached dwellings produce more electricity than they consume.[4] For example, in the Kakadu case study, the energy infrastructure was partly funded by the Federal Department of Families, Housing, Community Services and Indigenous Affairs. In such instances, a comprehensive approach to the housing needs should be considered when designing the solar infrastructure.

In order to attract funding, X Group should make the case that the project is sustainable in accordance with the SDP and SCIIs™. X Group should spot issues and resolve those issues and also point out the benefits to the funding body or private investor.

Issues and benefits under the SDP spotted in brief:

Technological sustainability

X Group should use some of the grant money to undertake due diligence on the kind of technology appropriate for the site. Most remote desert communities suit hybrid solar PV systems with battery storage or backup generators powered by diesel, waste plastic oil or biofuel.

If a hybrid solar PV system is selected, it will save on greenhouse emissions and on diesel fuel cost. The maintenance costs will be lower than a stand-alone diesel generator. The cost of transporting diesel will also be eliminated or greatly reduced.

Economic sustainability

In order to fund the ongoing maintenance and upgrades of the infrastructure and to fund new builds in the community, X Group should establish a trust fund. Savings from the cost of diesel, each quarter should be put into the trust fund for the purpose of upgrading and maintaining the solar panels and other infrastructure builds for the community such as street lights, electric appliances, an electric water pump, first aid station, community garden and school house with school of the air connectivity.

A proportion of other community income can also be placed into this trust fund for future infrastructure builds and maintenance.

For example:

1. Rent from community housing;
2. Tourist enterprises;
3. Market gardens; and
4. Maintenance and construction services.
   Often a community with reliable solar power will attract population from nearby communities. If the community grows, income can also be drawn from:
5. Grocery stores;
6. Renewable energy education centres;
7. Renewable energy training centres; and
8. Rent from properties where renewable energy infrastructure is built.

A business case on economic opportunity for the community should be made to the funding body.

Environmental sustainability

X Group should check the sustainability of the solar supplier, including information about cashback options for recycling batteries and panels at the end of the life cycle. The savings on greenhouse gas emissions and pollution from diesel emissions or the use of a recycled carbon fuel in a hybrid system will give the project a good sustainability rating. If housing is built, energy efficient architecture should be considered.

Sustainable governance

X Group does not have community members trained in solar PV maintenance. They do not have a close working relationship with neighbouring Y Group. Power struggles between neighbouring communities can often be overcome through joint venture agreements. Alternatively, some solar providers provide a remote maintenance service which entails initial training for community members, a detailed user manual and ongoing phone support to talk users through the user manual when performing maintenance. Additionally, a community member could be interested in travelling to Alice Springs to undergo formal training or could complete an online training course.

The social benefits to the project should be outlined in the grant application or investor brief showing the potential increase in standards of living.

- Reliable affordable electricity;
- Streetlights;
- Households can be productive in the evenings, for example, studying;
- Hygiene and health can improve through the use of electric appliances such as washing machines and fridges;
- Communication can be improved through television, mobile phones and radios;
- Health issues associated with diesel pollution are eliminated;
- Access to freshwater is made possible (ability to electrically pump bore water or store and pump rain water or refrigerate heated tap water in desert communities);
- Access to education such as school of the air or renewable energy education centres has increased;
- Construction of health clinics in remote communities could be enabled by the power source;
- Possible creation of jobs in electrical maintenance, construction, tourism and community farming;
- Access to reliable refrigeration and fresh food;
- Reduced amount of diesel fuel needed for generators;
- Improved employment and training opportunities; and
- Population could increase, indigenous people moved from isolated outlying areas into communities with solar power.

Tentative conclusion

Overall, it is possible for X Group to arrange their own renewable energy supply without relying on leasing infrastructure from Y Group. The above issues will need to be resolved to ensure sustainability of the project. Due diligence of the solar supplier, site suitability and advice from legal and accounting experts is also recommended.

Notes

1 Based on the New South Wales town of Uralla, that saw a property price rise of 35% in 2008 (the fastest property price rise in Australia according to Australian Property Monitors, May 15, 2008). Uralla has developed Australia's first zero-net energy community plan.
2 The National average.
3 Contact the National Indigenous Australians Agency for more information.
4 Sekisui House. 2018.

10

CONCLUSION

The current trend towards social and sustainable investing in industries such as renewable energy is attracting many market participants. From large superannuation and pension funds and green bond markets to local community investors. Many of the successful case studies demonstrated partnerships between government, private and community investors and participants.

Non-network solutions worked well in off-grid remote locations and large-scale solar was successful in remote on-grid agricultural communities. These large projects were profitable and provided renewable energy for locals as well as exporting energy to urban infrastructure such as the Sydney Opera House and Coles supermarkets.

In this time of productivity and growth, there is an opportunity to build Sustainable Systems, communities and industries. We can ask questions as to the levels of sustainability, such as are the funds raised in a sustainable way, for example, are loans serviceable? Is the technology sustainable in that the funds will be available to maintain, upgrade or decommission the project? This will depend on whether the governance structure allows for the saving or raising of capital throughout the life of the project such as in a joint stock corporation. Is the project attracting population migration to isolated regions? Is the project increasing the economic output of the region and are the standards of living improving as the population increases? Does the technology help to maintain the natural capital as population increases, for example, using solar instead of burning old growth forest for fuel? Is the transition responsible in that the energy access and affordability improves because of the project? Have social benefits and increased standard of living ensued such as the reduction of energy poverty and the access to employment and education? Are the communities' needs for democratic and human rights being upheld?

One of the fundamental aspects of democracy on which our society relies is the legal principle that constituents have the freedom to choose how to invest private

DOI: 10.4324/9781003324669-10

property and the freedom to alienate private property. It is this freedom that should be kept at the forefront of the mind of any well-meaning policy maker that might seek to draft regulations in relation to environmental or sustainable investment choices and strategies. As put eloquently by Rose,

> private (sic) property regimes generally mediate issues of resource use and discourage feuds while encouraging trades instead. As to trade, people meet others in market relationships; they learn to trust one another and to behave in trustworthy ways, and out of those relationships of trust they can develop general habits of civility and more specific friendships, sometimes quite remarkable ones. Property accepts people as they generally are—self-interested, to be sure, but capable of cooperation—and of course it leverages both traits into productive activity ....By contrast, it is the Utopian, first-best demand—the demand that insists on sharing and that concomitantly severely limits property—.... As a result, although there are some utopian successes such as monasteries, the history of utopian experiments is littered with moral failure and sometimes great cruelty.[1]

The strength of the Sustainable Community Investment Indicators (SCIIs™) decision-making tool is that it is based on qualitative research and context-specific case studies as opposed to quantitative research or statistical based research. This process has ensured that the SCIIs™ can be applied in various contexts and situations and tailored for the local communities in which they are applied. The Sustainable Development Proposition (SDP) working in conjunction with the SCIIs™ provides an incentive-based tool that encourages best practice investment. The legal structure chosen by the community or investment project will of course vary widely, as has been demonstrated by the case studies.

A common sense approach to a global effort towards sustainability will necessarily be an approach that remains flexible such that the local structures and social needs of widely varying communities can be met. In this way, the SDP and the SCIIs™ can be applied to a limitless number of innovative and locally specific governance and social structures and needs. For example, in the case studies, some communities demonstrated a sustainable governance structure under an Aboriginal and Torres Strait Islander corporation, while other communities demonstrated success under a revolving fund structure. An innovative hybrid structure of an Aboriginal and Torres Strait Islander Revolving Fund is a possible governance structure. However, it would be nonsensical to then say that the hybrid Aboriginal and Torres Strait Islander Revolving Fund governance model will be the best practice and most sustainable model for every indigenous community. To impose such a rule would severely impinge on the democratic rights of those communities to invest and deal with property freely. The property investment needs of indigenous and rural communities will differ and be context specific.

Additionally, although several examples of technology that would be considered sustainable from the perspective of small regional communities (such as small-scale solar) based on the successful application of those technologies in the case studies analysed, the SCIIs™ go beyond that analysis and allow a decision-making tool that is flexible as technology advances. The strength of the SCIIs™ method is that with each new investment or project a new analysis is done in relation to the sustainability of the technology. This means that as technology advances and innovations occur, the newer and more sustainable inventions will naturally replace the outdated technologies previously used in older community projects. It is only under these adaptable market conditions that the Cobb-Douglas function explained above can operate such that it will ensure sustainability of renewable resources in the face of a growing population while supporting an improved standard of living. For this reason, the SCIIs'™ table does not have a list of example technologies in its matrix, whether that be financial, legal or renewable energy technologies that have been explored in the case studies. Best practice drafting of principles has been followed. Best practice drafting dictates that examples should be used sparingly, if at all, when drafting principles.

Furthermore, the SCIIs™ and SDP are designed to work in free-market conditions as opposed to public regulation conditions. There are many common sense reasons in a democratic environment to design investment decision-making tools in this way such as the encouragement of investment and innovation. The most important common sense reason is that it removes the breeding ground for corruption that arises in the "burgeoning use of property-like institutions in public regulation." Academics such as Sandel argue against the dangers of commodification and "that these undermine civic consciousness."[2]

The case studies demonstrate the most sustainable technological, environmental, governance and social aspects of the investments in those case studies. The advantage of the SDP and the SCIIs™ is that they have been designed to go beyond those recommendations and to be flexible enough to be applied in any community context going forward into the future.

Given the right input of data from experts in the field of law, planning, engineering and financial advice as it applies to a particular community, the sustainability of that future project can be analysed and promoted.

Another advantage of the SDP and the SCIIs™ is that although they are designed to be flexible and work across a variety of community contexts, they can also be applied to a more rigid and public regulated system such as the European Commission's Taxonomy Regulation and the Sustainable Finance Disclosure Regulation.[3] This system has been criticised for its rigidity and unclear drafting and has created concerns that its design will actually impede environmental and sustainable governance investing.[4] On the other hand, a company that can manage this new and complex regulatory environment can distinguish itself in the marketplace and gain a competitive advantage.[5] For instance, the Lighthouse Infrastructure Investment Management Company has used the regulation to distinguish itself in the Queensland solar industry.[6]

Due to the flexibility of its design, SCIIs™ can be used by a financial investor when navigating formal regulated systems such as the Sustainable Finance Disclosure

Regulation and other regulations being produced under the European Green Deal. In fact, they can be applied to any regulatory context as the analysis of the governance systems in the SCIIs™ matrix is done in a context-specific method rather than a fixed regulatory method and so can be applied to analyse any governance structure.

The spirit of this design is democratic, which requires that alternative policy options must be provided to communities and nations and their policy makers as policy options.[7] It is this vein that the research into guidance for communities has been conducted, to provide options.

Guidance must be provided in a transparent and understandable way such that local decision-makers can examine the trade-offs against the impacts across a decision-making matrix and decide for their own communities how to improve the investment proposals so as to maximise opportunities for a win-win outcome for their communities.[8] There are different tools that can be used to provide this sustainability analysis to communities.

The SDP working in conjunction with the SCIIs™ has been designed to allow for diversity across communities in rural and regional areas and as such is a flexible matrix. The by-product of this flexibility is that it can be applied across any number of regulated and unregulated markets and legal systems. This decision-making tool is not a "how to guide" for practitioners or experts but is a tool to form the basis of a more tailored plan for investment in a context-specific project. Democratic countries use a variety of methodologies and tools in running sustainability assessments. The SCIIs™, in the spirit of democracy, are tools to assess the viability of an investment from a sustainability perspective for remote communities. The choices made resulting from that analysis will depend on a variety of context-specific factors, such as, the availability of resources, institutional capacities and legal structure choices made at governance levels in a variety of communities.

The SDP and the SCIIs™ also operate as an investment incentive tool in the spirit of free-market trade. They provide best practice analysis to allow a project to promote themselves as attractive for sustainably minded investors. This model is to be distinguished from a risk assessment model which is used by quantity surveyors, insurance companies or lawyers to advise clients on how to avoid risk such as potential law suits or over spending. For example, the Principals for Responsible Investment Sustainable Development Goals[9] that have informed the development of the European Commission's Taxonomy Regulation and the Sustainable Finance Disclosure Regulation are stated as being a "risk framework" and "capital allocation guide." This kind of framework is useful in the due diligence stage of a project; however, it is quite different from a sustainability assessment that is aimed at attracting investment and encouraging economic growth and infrastructure investment in communities that would be otherwise overlooked. In fact, risk assessment based tools are more likely to hinder than promote investment as investors must avoid risky investments such as small communities without a financial track record, in order to comply with the rules.

Formulating an assessment based on avoiding legal and environmental risk will not necessarily attract investment even where the risks analysed are sustainability risks. Risk assessments and due diligence are required prior to the construction phase of an infrastructure project. The SCIIs™ are designed to attract investment by promoting the sustainable aspects of a potential investment and by monitoring the success of those investments, with the aim of improving them for the subsequent attraction of further investment. This is to be distinguished from the aim of an insurer whose objective is to avoid risk and who does not have a goal of attracting investment but has a different goal of reducing the risk of investments.

A by-product of the SCIIs™, however, is that by helping to design a system for a community that is more sustainable, it follows that that system will be less exposed to the risks that come from unsustainable investments and therefore the SCIIs™ will assist with compliance within a regulatory system that is based on "risk frameworks." The more flexible those risk frameworks are the better it will be for small communities who might have a high risk profile for investment.

The SCIIs™ have been designed to create this value proposition for investors and can be used to inform such sustainability analyses that are being taken up by companies globally. When adapted flexibly to the small context-specific community project, this may well attract investment to such a small community from a larger investment company with sustainability goals.

The SCIIs™, a unique community investment indicator matrix, as well as an original SDP can be used to attract investment in renewable energy projects for remote and indigenous communities.

This cost-benefit analysis of multiple renewable energy projects in rural and indigenous communities to investigate the financial, legal and technological structures in community case studies analysed funding models for sustainability and their attractiveness to investors. The most attractive and sustainable investment strategies were distinguished, selected and recommended. Novel hybrid structures developed out of this analysis and were identified.

Several layers of analysis have been conducted and the research questions have been answered. First, the current state of energy infrastructure in Australian rural and Indigenous communities and found a gap in infrastructure investment thus confirming the significance of undertaking this research. Next, a property economics approach was evaluated and found to be a valid investigative tool.

The current energy governance system was analysed and it was found that a free-market investment approach was available under the current system and that decentralised energy islands were supported and recommended by the Chief Economist.

This paved the way for evaluating models within community case studies that could overcome any limitations to remote energy supply using a cost-benefit analysis to differentiate and select recommendations.

Finally, a SDP and the Sustainable Community Investment Indicators were developed to assist with decision-making, implementing and monitoring those recommendations. The SDP was informed by the case study analysis as well as a literature review of the concept of sustainability. Within that context, an analysis of currently available sustainable energy technology and infrastructure for remote communities was also conducted. A review of sustainable governance methods and sustainable economic methods was conducted. This literature review in conjunction with the case study analysis was synthesised and thus informed the development of the Sustainable Community Investment Indicators.

The advantages of this decision-making tool are that they can be implemented across a variety of community projects and can be adapted to the particular community context. This is important because what is sustainable for one community could be unsustainable for another. It is imperative that the self-determination of the community decision-makers remains democratic and free to embed their design choices, technology choices and legal structure choices that will be the most sustainable for their own particular community and local environment. For instance, the micro hydro technology will not be sustainable in a drought area and the small-scale solar will not be suitable for an area with protected forest and extensive shade coverage. The various corporate structures were shown to be advantageous in different kinds of communities and contexts but disadvantageous in others. The SCIIs™ encourage sustained, multi-layered engagement from community decision-makers without imposing external, centralised control or regulation that could impair that self-determination and autonomy and democratic freedom to invest.

The detailed case study analysis demonstrated the importance of context-specific enterprise design on community autonomy and sustainability of projects. Participation by communities needs to be emergent and adaptable in order to generate participation that is significant, sustained and diverse. The diversity of choice within the free market is what attracts the vast range of investors in community projects in western countries. The SCIIs™ specifically do not list too many examples of technology or other sustainable practices as this leaves room for innovation and context-specific adaptation to self-determinative community investment choices.

The motivation to investment comes from context-specific community needs and wants and relies on the flexibility of the investment market to adapt to provide where there is a gap in different community contexts. This is the primary advantage that was found in the free-market investment solutions over and above the controlled regulatory or grant-driven solutions.

Each case study explored context-specific engagement, economic arrangements, legal structure practices and technology choices and sustainable outcomes related to those context-specific choices. Community participation in enterprise design increased local investment and involvement, creation of local relationships and employment and increased skills in the communities. A range of social, economic and environmental impacts were reviewed across a variety of diverse community projects.

Although not every project will or should be expected to achieve sustainability across all of the SCIIs™, given the variety of community contexts, the design choices of the project can be informed by this tool and it can lead to better design choices or best practice design choices. Participation and engagement by the local community will be a factor beyond the control of many large investors unless a subjective, context-based approach to community design involvement is conducted when using the SCIIs™. Depending on the project, multiple parties could be involved in the pre-construction phase and procuring investment process. The SCIIs™ are flexible such that communities can prioritise SCIIs™ according to their own unique enterprise design.

Community empowerment comes not only from income or job creation but also from involvement in the design of the project and the SCIIs™ analysis can be done from a community context-specific perspective. It is for these reasons that the SCIIs™ are designed as a decision-making tool. The advantage of this being the creation of community engagement and participation. The case studies have shown that mere investment of money without community centred design involvement will not necessarily produce sustainable outcomes for community projects. However, thoughtful application of best practice project design can attract sustainable investment and project longevity with positive community outcomes.

Many community organisations live from hand-to-mouth, spend lots of time trying to entice people to donate often small sums and come very low in the pecking order of government expenditure. They often campaign for more public expenditure without the knowledge of financial and legal innovation that would help them avoid this queue for funding. Likewise, "shaking a collection tin in a local town centre is worthy, but usually the takings do not remotely match the needs.

FIGURE 10.1 An artistic photo of a light bulb on the forest floor with a seedling growing inside it. Photo by Singkham from Pexels.

It is easy to be dispirited."[10] Additionally, large loans that would be used to secure funds to build infrastructure are difficult to secure or service for many communities. Furthermore, forcibly attempting to procure investment from the financial services industry through an Environmental Sustainable Governance (ESG)code is the antitheses of the free market and democratic decision-making process that attracts investment from free-market investors and can have adverse impacts on social investing in small communities that are considered high risk by institutional investors.

The property economics approach can be used to address lack of infrastructure where grid connection is untenable due to large distances being covered. Attracting investment into community development is something which requires broad and detailed analysis. It is not sufficient, for example, to merely rely on normative processes and census data to research the needs of the community and from here provide adequate grant funded infrastructure for the community to increase the standard of living.[11] Neither is it adequate to rely upon isolated, community-orientated data collection and consultation.[12] Sustainable community development needs precise analysis based on processes, policies, principles and practices. It requires a well-rounded analysis of the legal, social, political, economic and environmental impacts. There is a high level of responsibility when developing communities. Fundamental rules of law and economic principles must not be left out of the analysis due to a narrow focus on one factor, such as, engineering, town planning, social or economic benefit or environmental impact. It is from this qualitative research basis that recommendations can be made.

By following the SDP communities can attract investment and sustain that investment model to keep pace with other communities and provide a legacy for future generations.

As recognised in the *Special IPCC Report*, innovative financing mechanisms are needed to help create sustainable development practices. They can draw on bottom-up approaches and use indigenous knowledge to effectively engage and protect vulnerable people and communities in the face of climate change vulnerability and lack of essential infrastructure.[13]

As demonstrated by the case studies, small-scale solar projects are highly cost-effective with a payback period of two to three years. When approaching investors in the pre-development stage of an infrastructure project, the SDP and Sustainable Community Investment Indicators (SCIIs™) can help articulate the benefits of the renewable energy infrastructure. The team of professionals assessing the viability of the development can use the SCIIs™ matrix to flesh out a complete proposal that is attractive for investors.

In the case studies evaluated many of the remote community projects were driven by local, small-scale community investors. This category of investor is motivated and attracted by social and environmental benefits for the community as well as autonomous and transparent legal governance structures. The financial structure is also crucial to attracting local small-scale investment. Even where the profits or dividends are moderate, when a project can demonstrate that there is a return on investment and that there is a short payback period on the CAPEX, local community investors are attracted to the project.

Despite the relatively moderate profits in small-scale projects, many large and sophisticated investors were involved in investing in the projects outlined in the case studies. They were motivated, not only by returns on their investments, but also by the social and environmental gains for the communities. Additionally, many large investors found novel ways to scale up investment in small infrastructure projects in remote communities. Publicly listed companies and superannuation funds are under increasing pressure from shareholders to invest in renewable energy.[14] This creates an opportunity to attract investment, even in small-scale projects, from large investors. This would especially be the case where a series of small-scale projects were linked together as in the Grameen Bank, CfAT and CORENA case studies. Additionally, sustainable bonds, super funds and other kinds of publicly listed funds are now increasing on the open stock market.[15] These kinds of funds require sustainable projects for their investment portfolios. The SDP and SCIIs™ can be used to assess and attract investment from these large investors.

The case studies have demonstrated that both local community investors and large corporate or government investors can realise profits by investing in renewable energy projects in remote communities. The SDP and SCIIs™ can be used to tailor an investment proposal to different categories of investor, highlighting the strengths and benefits of a community project investment. The SCIIs™ can also be used to monitor the success of an investment in the post-construction phase.

This work is significant not only because there is currently a gap in energy security for remote communities but also because of the opportunities that addressing this gap can provide for those communities. By creating a sustainable infrastructure project, local employment opportunities, economic growth and contribution to a more sustainable management of resources can be achieved by communities and their investor partners.

This guide challenges the current reliance on traditional funding methods of publicly funded government grants and private institutional investment and recognises that while a government regulatory system is needed in the electricity marketplace, that system supports an economically diverse plethora of sustainable investment models to meet the needs of the diverse communities in rural and indigenous communities.

Notes

1 Rose 2007, 1987, 1929.
2 Ibid. 122–23.
3 European Commission, Sustainable Finance Overview.
4 Dewi 2021.
5 KPMG 2021.
6 Principles for Responsible Investment 2021.

7 The Organisation for Economic Co-operation and Development (OECD) "Guidance on Sustainability Impact Assessment".
8 Sovacool et al. 2014. *Energy Research & Social Science.* 14 onwards.
9 United Nations "Sustainable Development Goals", 2021.
10 Helm 2015, 13.
11 Kenny, S and Connors 2016, 396.
12 Ibid. 397.
13 The United Nations, The Intergovernmental Panel on Climate Change, *Special Report*, 2018, 40.
14 Klimenko 2019.
15 Potter 2021.

REFERENCES

Aboriginal agl. 2008. Aboriginal and Torres Strait Islander Land Tribunals annual report 2007–2008. www.courts.qld.gov.au/__data/assets/pdf_file/0010/269353/atsi-land-tribunal-annual-report-2007-08.pdf

Aboriginal Land Tribunal Queensland. 2019. Land Tribunal annual report 2018–2019. www.courts.qld.gov.au/__data/assets/pdf_file/0008/635633/land-tribunal-annual-report-2018-19.pdf

AGL. n.d. "Get more from your solar battery." Accessed February 7, 2020. www.agl.com.au/solar-renewables/solar-energy/bring-your-own-battery

Andersson, Krister P. and Elinor Ostrom. 2008. "Analyzing decentralized resource regimes from a polycentric perspective." *Policy Sciences* 41: 71–93. https://doi.org/10.1007/s11077-007-9055-6

Andhina, Putri and P. Purwanto. 2018. "Evaluation of livestock waste management to energy biogas (Case study: Jetak Village, Getasan Sub District)." *E3S Web of Conferences* 73: 1–2. https://doi.org/10.1051/e3sconf/20187307013

Australian Bureau of Statistics. "Births, Australia" www.abs.gov.au/statistics/people/population/births-australia/latest-release

Australian Bureau of Statistics. 2017. "Census of Population and Housing: Reflecting Australia – Stories from the Census 2016." Released June 28, 2017. www.abs.gov.au/ausstats/abs@.nsf/mf/2071.0

Australian Bureau of Statistics. 2002. "Community housing and infrastructure needs survey 2001." Released June 6, 2002. www.abs.gov.au/AUSSTATS/abs@.nsf/33a109e2dcda21dfca2570920021012a/b65ab82e947fcbd4ca256bd0002827f0!OpenDocument

Australian Bureau of Statistics. 2007. "Community housing and infrastructure needs survey. Australia, Data dictionary, 2006 (Reissue)." Released August 23, 2007. www.abs.gov.au/AUSSTATS/abs@.nsf/Lookup/4710.0.55.001Main+Features12006%20(Reissue)?OpenDocument

Australian Bureau of Statistics. 2019. Environmental Assets. Accessed February 3, 2020. www.abs.gov.au/AUSSTATS/abs@.nsf/allprimarymainfeatures/9EF05B385442E385CA257CAE000ED150?opendocument

Australian Bureau of Statistics. 2008. "Australian Social Trends: Housing and services in remote Aboriginal and Torres Strait Islander communities." Released July 23, 2008. www.abs.gov.au/AUSSTATS/abs@.nsf/Lookup/4102.0Chapter9202008

Australian Competition and Consumer Commission. 2018. Restoring electricity affordability and Australia's competitive advantage: Retail electricity pricing inquiry – final report. www.accc.gov.au/publications/restoring-electricity-affordability-australias-competitive-advantage

Australian Competition and Consumer Commission. 2019. Consumer data right in energy: Consultation paper: Data access models for energy data. www.accc.gov.au/focus-areas/consumer-data-right-cdr/cdr-in-the-energy-sector/consultation-on-energy-data-access-models

Australian Energy Market Commission. 2017. Alternatives to grid-supplied network services. www.aemc.gov.au/Rule-Changes/Alternatives-to-grid-supplied-network-services

Australian Energy Market Commission. 2019. How digitalisation is changing the NEM: The potential to move to a two-sided market. www.aemc.gov.au/news-centre/corporate-publications/how-digitalisation-changing-nem

Australian Energy Market Commission. n.d. "National Electricity Rules." Accessed July 18, 2019. www.aemc.gov.au/regulation/energy-rules/national-electricity-rules

Australian Energy Market Operator. 2017. Visibility of distributed energy resources: Future power system security program. www.aemo.com.au/-/media/Files/Electricity/NEM/Security_and_Reliability/Reports/2016/AEMO-FPSS-program----Visibility-of-DER.pdf

Australian Energy Market Operator. n.d. "About AEMO." Accessed July 18, 2019. www.aemo.com.au/About-AEMO

Australian Energy Regulator. 2009. State of the energy market report. www.aer.gov.au/publications/state-of-the-energy-market-reports/state-of-the-energy-market-2009

Australian Energy Regulator. 2020. Annual Report.

Australian Government. 2012. Department of Social Services Remote Indigenous Energy Program Questions and Answers, 25 June 2012 at www.dss.gov.au/sites/default/files/documents/07_2012/remote_indigenous_energy_program_qas_0.pdf

Australian Government. 2013. Land Use – Australia. Australian Natural Resource Atlas.

Australian Government. "Funding to support studies into improved energy supply arrangements." Accessed March 2, 2021. www.business.gov.au/grants-and-programs/regional-and-remote-communities-reliability-fund-microgrids

Australian Government. "Minister Frydenberg: Delivering solar energy to remote Indigenous communities at www.indigenous.gov.au/news-and-media/announcements/minister-frydenberg-delivering-solar-energy-remote-indigenous 13 May 22.

Australian Government, Department of Agriculture, Water and the Environment. 2020. "Protected Matters Search Tool." Accessed February 3, 2020. www.environment.gov.au/webgis-framework/apps/pmst/pmst.jsf

Australian Government, Department of the Environment and Energy. 2015. Australia's "True Up Period" Report for the Kyoto Protocol First Commitment Period. www.industry.gov.au/data-and-publications/australias-true-up-period-report-for-the-kyoto-protocol-first-commitment-period

Australian Government, National Indigenous Australians Agency. "Review of the CATSI Act." Accessed March 3, 2021. www.niaa.gov.au/indigenous-affairs/economic-development/review-catsi-act

Australian Government, Office of the Registrar of Indigenous Corporations. 2010. What the CATSI Act means for funding bodies. www.oric.gov.au/publications/catsi-fact-sheet/what-catsi-act-means-funding-bodies

Australian Government. Department of the Environment. 2015. National Pollutant Inventory Emission data.

Australian Medical Association. 2013. Submission to Senate Community Affairs References Committee, Parliament of Australia. Impacts on Health of air quality in Australia.

Australian Renewable Energy Agency. 2020. "Integrating concentrating solar thermal energy." Accessed May 22, 2020. https://arena.gov.au/projects/integrating-concentrating-solar-thermal-energy-into-the-bayer-alumina-process/

Australian Renewable Energy Agency. 2020. "Australian renewable energy mapping infrastructure." Accessed February 3, 2020. https://nationalmap.gov.au/renewables/

Australian Renewable Energy Agency. 2019. "Biogas Opportunities for Australia," ENEA Consulting and Bioenergy Australia.

Australian Renewable Energy Agency. 2020. "Solar power in Australia." Accessed May 14, 2020. https://arena.gov.au/renewable-energy/solar/

Australian Treasury. 2019. Privacy Impact Assessment: Consumer Data Right: March 2019. https://treasury.gov.au/publication/p2019-t361555

Axup, Kate. 2017. "Finkel Review – Renewables: The importance of regulatory certainty for renewables." www.allens.com.au/insights-news/insights/2017/06/finkel-review---renewables-the-importance-of-regulatory/

Barber, Gregory. 2021. "Sodium Batteries May Power Your New Electric Car." www.wired.com/story/sodium-batteries-power-new-electric-car/

Barrett, Adam. 2018. "Stability of zero-growth economics analysed with a Minskyan model." *Ecological Economics* 146 (C): 228–239. https://doi.org/10.1016/j.ecolecon.2017.10.014

Bartsch, Hans-Jochen. 2015. *Handbook of Mathematical Formulas*. Kent: Elsevier Science.

Bates, Gerard Maxwell. 2016. *Environmental law in Australia*. 9th ed. Chatswood: LexisNexis Butterworths.

Bates, Gerard Maxwell. 2019. *Environmental law in Australia*. 10th ed. Chatswood: LexisNexis Butterworths.

Beyond Zero Emissions and MRC agency. 2016. Zero Carbon Australia: Electric vehicles. https://bze.org.au/research/transport/electric-vehicles/

Beyond Zero Emissions and University of Melbourne Energy Research Institute. 2010. Australian sustainable energy: Zero Carbon Australia Stationary Energy Plan. http://media.bze.org.au/ZCA2020_Stationary_Energy_Report_v1.pdf

Beyond Zero Emissions. 2015. *The energy freedom home*. Melbourne: Scribe Publications.

Beyond Zero Emissions. 2018. Zero Carbon industry plan: Electrifying industry. https://bze.org.au/wp-content/uploads/electrifying-industry-bze-report-2018.pdf

Beyond Zero Emissions. 2019. How to electrify plastic. https://bze.org.au/wp-content/uploads/BZE-Electrifying-Plastic-Brief-2019-Beyond-Zero-Emissions-Australia.pdf

Bird, Robert, Baum, Zachary J., Xiang Yu, and Jia Ma. 2022. "The Regulatory Environment for Lithium-Ion Battery Recycling," *ACS Energy Letters* 7 (2): 736–740. doi: 10.1021/acsenergylett.1c02724

Bomen Solar Farm. Spark Renewables at www.bomensolarfarm.com.au/Accessed June 10, 2022.

Brown Weiss, Edith. 1989. *In fairness to future generations*. Tokyo: United Nations University; New York: Transnational Publishers.

Buergelt, Petra T., Elaine L. Maypilama, Julia Mcphee, Galathi Dhurrkay, Shirley Nirrpuranydji, Sylvia Mänydjurrpuy, Marrayurra Wunungmurra, Timophy Skinner, Anne Lowell, and Simon Moss. 2017. "Housing and overcrowding in remote Indigenous communities: Impacts and solutions from a holistic perspective." *Energy Procedia* 121: 270–277. https://doi.org/10.1016/j.egypro.2017.08.027

Bulletpoint. 2012. "The Remote Indigenous Energy Program." Accessed April 4, 2019. www.bulletpoint.com.au/remote-Indigenous-energy-program

Bushlight case studies. 2014. Accessed April 4, 2019. https://cfat.org.au/bushlight-archive/

Business Queensland. 2016. "Biogas production." Accessed May 7, 2020. www.business.qld.gov.au/industries/mining-energy-water/energy/renewable/projects-queensland/starting-biogas-project/biogas-production

Canberra Institute of Technology. "Renewable Energy Skills – Centre for Excellence" at https://cit.edu.au/industry_business/customised_training/renewable_energy_skills_training_centre

Centre for Air Quality & Health Research and Evaluation. 2013. Submission no 29 to Senate Community Affairs References Committee, Parliament of Australia, Impacts on Health of Air Quality in Australia.

Centre for Appropriate Technology (CfAT). 2000. Renewable Energy in Remote Australian Communities (Market Survey) at https://cfat.org.au/energy-resources

Centre for Appropriate Technology. 2018. Annual report 2017–2018. https://cfat.org.au/s/CATLtd-201718-Annual-Report-WEB-rp5p.pdf

Centre for Appropriate Technology. 2014. "Bushlight energy archive." Accessed April 4, 2019. https://cfat.org.au/bushlight-archive/

Centre for Appropriate Technology. 2016. The Northern Territory Homelands and Outstations assets and access review: Final report. https://static1.squarespace.com/static/5450868fe4b09b217330bb42/t/57f6f64746c3c4ab7345af96/1475802710933/Final-Master-HOAAR-Report-Oct2016-1.pdf

Centre for Social Impact, First Nations Foundation and NAB. 2019. "Money stories: Financial resilience among Aboriginal and Torres Strait Islander Australians" (2019) at www.csi.edu.au/media/NAB_IFR_FINAL_May_2019_web.pdf

Chamber of Minerals and Energy of Western Australia (CME). "Lithium and Battery Minerals." Accessed March 2, 2021. https://cmewa.com.au/about/wa-resources/battery-minerals/

Chambers, Matthew. 2015. "AGL Energy turns its back on coal-fired power stations." The Australian, April 17, 2015.

Cherbourg Aboriginal Shire Council. "Housing Policy," Accessed March 3, 2021. https://cherbourg.qld.gov.au/wp-content/uploads/2020/05/Housing-Policy.pdf

Chisholm, Richard and Garth Nettheim. 2002. *Understanding law: An introduction to Australia's legal system*. 6th ed. Sydney: LexisNexis Butterworths.

Citizens Own Renewable Energy Network Australia. n.d. (CORENA) "About." Accessed May 8, 2020. https://corenafund.org.au/what-we-do/

Citizens Own Renewable Energy Network Australia. n.d. "Quick win projects." Accessed May 8, 2020. http://corenafund.org.au/quick-win-projects

City of Darebin. n.d. "Energy and climate." Accessed May 11, 2020. www.darebin.vic.gov.au/en/Darebin-Living/Caring-for-the-environment/EnergyClimate

City of Adelaide. n.d. "Solar Savers." Accessed May 11, 2020. www.cityofadelaide.com.au/your-council/funding/solar-savers-adelaide

Clark, Eric. 2018. An overview of the Photovoltaic and Electrochemical Systems Branch at the NASA Glenn Research Center. https://ntrs.nasa.gov/citations/20190001013%202019-11-05T04:26:12%2000:00Z

Closing the Gap Implementation Plan. 5th August 2021: See also Council of Australian Governments. 2017. National Indigenous Reform Agreement (Closing the Gap). www.federalfinancialrelations.gov.au/content/npa/health/_archive/indigenous-reform/national-agreement_sept_12.pdf

COAG Energy Council, Energy Security Board. 2019. Post 2025 market design for the National Electricity Market (NEM). www.coagenergycouncil.gov.au/publications/post-2025-market-design-national-electricity-market-nem

COAG. 2013. *Revised National Principles for Feed-In-Tariff Arrangements*. Council of Australian Governments.

Coalition for Community Energy. 2015. National community energy strategy. http://c4ce.net.au/nces/wp-content/uploads/2015/04/NCES_2015_Final01.pdf

Coalition for Community Energy. 2017. "Small-scale community solar guide webinar." Accessed April 21, 2020. https://c4ce.net.au/strategic-initiatives/webinars-2017-small-scale-community-solar-guide/#introduction

Coalition for Community Energy. 2017. Small-scale community solar guide. http://c4ce.net.au/wp-content/uploads/2017/09/C4CE-Small-Scale-Community-Solar-Guide-v2.pdf

Coalition for Community Energy. 2018. "C4CE webinar 6: Revolving funds rock with CORENA and COREM."Youtube video, November 6, 2018. https://youtu.be/ZYWaHGGNidc

Coalition for Community Energy. 2018. "C4CE webinar 7: Upper Yarra community power." Youtube video, November 13, 2018. www.youtube.com/watch?v=Y2zu85e-jeQ

Collyer, Anna, Rosannah Healy and Robert Walker. 2018. *National energy guarantee on hold in favour of interventionist powers*. www.allens.com.au/insights-news/insights/2018/08/national-energy-guarantee-on-hold-in-favour-of/

Common Capital. 2015. "Community Solar Financing: Research Summary" at https://commoncapital.com.au/

Common, Michael S. and Sigrid Stagl. 2005. *Ecological economics: An introduction*. Cambridge: Cambridge University Press.

Commonwealth Scientific and Industrial Research Organisation. 2019. Artificial Intelligence: Australia's ethics framework: A discussion paper. https://consult.industry.gov.au/strategic-policy/artificial-intelligence-ethics-framework/

Copper, Jessie K., Alistair B. Sproul and Stefan Jarnason. 2016. "Photovoltaic (PV) performance modelling in the absence of onsite measured plane of array irradiance (POA) and module temperature." *Renewable Energy* 86: 760–769. https://doi.org/10.1016/j.renene.2015.09.005

Council of Australian Governments Energy Council, Energy Security Board. 2019. Post 2025 market design for the National Electricity Market (NEM) www.coagenergycouncil.gov.au/publications/post-2025-market-design-national-electricity-market-nem

Council of Australian Governments Energy Council. 2016. Stand-alone energy systems in the Electricity Market: Consultation on regulatory implications. http://coagenergycouncil.gov.au/sites/prod.energycouncil/files/publications/documents/Stand-Alone%20Energy%20Systems%20Consultation%20Paper%20-%20August%202016.pdf

CSIRO, GenCost. 2019–2020. Preliminary results for stakeholder review. Draft for review, December 2019.

Damodharan, Dillikannan, Babu Rajesh Kumar, Kaliyaperumal Gopal, Melvin Victor De Poures and Balaji Sethuramasamyraja. 2019. Utilization of waste plastic oil in diesel engines: A review. *Reviews in Environmental Science and Biotechnology* 18 (4): 681–697. https://doi.org/10.1007/s11157-019-09516-x

Delucchi, Mark A. and Mark Z. Jacobson. 2011. "Providing all global energy with wind, water, and solar power, Part II: Reliability, system and transmission costs, and policies." *Energy Policy* 39 (3): 1170–2190. https://doi.org/10.1016/j.enpol.2010.11.045

Denniss, Richard. and Campbell, Rod. 2015. *Two Birds, one little black rock, solving the twin problems of incentives for retirement of coal fired generation and funding rehabilitation liabilities*. The Australia Institute.

Derwent, Richard, Peter Simmonds, Simon O'Doherty, Alistair Manning, William Collins and David Stevenson. 2006. "Global environmental impacts of the hydrogen economy." *International Journal of Nuclear Hydrogen Production and Application* 1 (1): 57–67. https://doi.org/10.1504/IJNHPA.2006.009869

Desert Knowledge Australia. 2019. Annual Report 2019. www.dka.com.au/about-dka

Dewi, John. 2021. "Europe's ESG regulations: Green Investing meets red tape." Accessed July 30, 2021. Refinitiv at www.refinitiv.com/perspectives/regulation-risk-compliance/europes-esg-regulations-green-investing-meets-red-tape/

Dickers, Jessica. 2020. *Non-network solutions needed to manage demand.* Utility Engineering Construction and Maintenance. June 16, 2016, at https://utilitymagazine.com.au/non-network-solutions-needed-to-manage-demand/ accessed 9 March 2020.

Dikotter, Frank. 2010. *Mao's Great Famine: The history of China's most devastating catastrophe, 1958–1962.* London: Bloomsbury.

Downes, Jacqueline. 2018. 'ACCC's plan to lower network costs for consumers.' Accessed February 7, 2020. www.allens.com.au/insights-news/insights/2018/08/acccs-plan-to-lower-network-costs-for-consumers/

Downes, Jacqueline. 2018. 'ACCC wants changes to the National Electricity Market' Accessed April 2020. www.allens.com.au/insights-news/insights/2018/08/accc-wants-changes-to-the-national-electricity-market/

Drahos, Peter and John Braithwaite. 2002. *Information feudalism: Who owns the knowledge economy?* Abingdon: Earthscan.

Droege, Peter. 2009. *100% renewable energy autonomy in action.* London: Earthscan.

Edis, Tristan. 2015. "Reality Check: Origin Energy's Green Commitments, Climate Spectator. October 23 2015.

Egender, Joe. and Leeor Kaufman. 2019. "Unnatural Selection". Netflix documentary.

Ehtiwesh, Ismael. et al. (2019) "Deployment of parabolic trough concentrated solar power plants in North Africa – a case study for Libya." *International journal of green energy. [Online]* 16 (1): 72–85.

Ekistica and Centre for Future Energy. 2020. "Off-Grid Guide." https://intyalheme.dka.com.au/projects/off-grid-guide

Elinor, Ostrom. 2010. "Polycentric systems for coping with collective action and global environmental change." *Global Environmental Change* 20 (4):550–557. https://doi.org/10.1016/j.gloenvcha.2010.07.004

Elinor, Ostrom. 1993. *Governing the commons: The evolution of institutions for collective action.* Cambridge: Cambridge University Press.

Energy Networks Australia and Commonwealth Scientific and Industrial Research Organisation. 2017. Electricity Network Transformation Roadmap: Final report. www.energynetworks.com.au/projects/electricity-network-transformation-roadmap/

Energy Networks Australia. 2017. Cyber security and energy networks. www.energynetworks.com.au/sites/default/files/16022017_cyber_security_and_energy_networks_a4.pdf

Energy Queensland. 2018. Energising Queensland communities: Annual Report 2017–2018. www.energyq.com.au/__data/assets/pdf_file/0004/691123/Annual-Report-COMPLETE.pdf

Energy Queensland. 2010. "Charter Report" at www.energyq.com.au/__data/assets/pdf_file/0011/784109/Energy-Charter-Report-2018-19.pdf

Engineers Australia. 2017. Independent review into the Future Security of the National Electricity Market. www.environment.gov.au/submissions/nem-review/engineers-australia.pdf

Environmental Defenders Office. 2019. "What is a DOGIT in Queensland?". www.edoqld.org.au/explainer_what_is_a_dogit

Eras-Almeida, Andrea. et al. 2019. "Lessons learned from rural electrification experiences with third generation solar home systems in latin America: Case studies in Peru, Mexico, and Bolivia." *Sustainability (Basel, Switzerland). [Online]* 11 (24): 7139.

Erickson, Jon D. and Duane Chapman. 1995. "Photovoltaic technology: markets, economics, and rural development." *World Development* 23 (7): 1129–2141. https://doi.org/10.1016/0305-750X(95)00033-9

Ernst & Young Global Limited. 2019. *Renewable Energy Country Attractiveness Index*. https://assets.ey.com/content/dam/ey-sites/ey-com/en_ro/news/2019/12/ey-recai-country-index-and-chart.pdf

European Commission. 2020. "Standards: City – Smart cities." Accessed May 12, 2020. https://ec.europa.eu/eip/ageing/standards/city/smart-cities_en

European Commission. Internal Market, Industry, Entrepreneurship and SMEs. 2019. "Circular Plastics Alliance" at https://ec.europa.eu/growth/industry/policy/circular-plastics-alliance_en

European Commission. 2019. Sustainable Finance Overview at https://ec.europa.eu/info/business-economy-euro/banking-and-finance/sustainable-finance/overview-sustainable-finance/platform-sustainable-finance_en; https://ec.europa.eu/sustainable-finance-taxonomy/

European Commission, Renewable Energy Directive (2018/2001/EU) at https://ec.europa.eu/energy/topics/renewable-energy/renewable-energy-directive/overview_en at 9th September 2020.

European Union. Circular Plastics Alliance Declaration, 20 September 2019, Brussels https://ec.europa.eu/growth/industry/policy/circular-plastics-alliance_en accessed January 8, 2020.

European Parliament. 2019. "European Green Deal Investment Plan". Accessed May 14, 2020. www.europarl.europa.eu/legislative-train/theme-a-european-green-deal/file-european-green-deal-investment-plan

Evans, Simon. 2021. "Emergency taskforce to fix diesel crisis", Australian Financial Review. The 9th of December 2021. At www.afr.com/companies/infrastructure/later-australia-s-diesel-engines-could-be-crippled-in-6-weeks-20211209-p59g6j

Eversole, Robyn. 2010. "Remaking Participation: Challenges for community development practice." *Community Development Journal* 47 (1): 29–41.

EWB Challenge. n.d. Kooma Traditional Owners Association, Australia. Accessed April 4, 2019. https://ewbchallenge.org/kooma-traditonal-owners-association-australia

Fairley, Peter. 2019. "China's Ambitious Plan to Build the World's Biggest Supergrid: A massive expansion leads to the first ultrahigh-voltage AC-DC power grid." https://spectrum.ieee.org/energy/the-smarter-grid/chinas-ambitious-plan-to-build-the-worlds-biggest-supergrid

Filatoff, Natalie. 2020. PV Magazine, "Australia is a key player in new Global Power System Transformation team of 6." 19 October 2020. www.pv-magazine-australia.com/2020/10/19/australia-a-key-player-in-new-global-power-system-transformation-team-of-6/

Finkel, Alan and Michael Graves. 2016. The future of renewables, http://allens.publish.viostream.com/simpletemplate?v=ota87gbxwmejq

Finkel, Alan. 2017. Independent review into the future security of the National Electricity Market: Blueprint for the future. www.energy.gov.au/publications/independent-review-future-security-national-electricity-market-blueprint-future

First Footprints. 2013. Film directed by Bentley Dean and Martin Butler. Voiceover by Ernie Dingo.

Fitzgerald, Anne M., Neale Hooper and Cheryl Foong. 2010. "Review of regulatory framework for environmental information." 2010 Sixth IEEE International Conference on e-Science Workshops. 2010: 154–159. https://doi.org/10.1109/eScienceW.2010.34

Fitzgerald, Bridget, Joanna Prendergast, Matt Brann, Tara De Landgrafft and Kit Mochan. 2017. "Powering the bush: Problems and solutions in rural Western Australia" Accessed May 22, 2020. www.abc.net.au/news/rural/2017-06-12/powering-the-bush-problems-and-solutions-in-western-australia/8598768

Foley & Lardner Group. Powering the future, "Commercial opportunities and legal developments across the EV batteries lifecycle." Accessed May 14, 2020. www.allens.com.au/globalassets/pdfs/campaigns/linklaters-powering-the-future.pdf

Fraas, Lewis M. 2014. *Low-cost solar electric power.* Cham: Springer International Publishing. https://doi.org/10.1007/978-3-319-07530-3

Fraunhofer ISE. 2020. "Recent Facts about Photovoltaics in Germany," Accessed May 20, 2020. www.ise.fraunhofer.de/en/publications/studies/recent-facts-about-pv-in-germany.html

Frontier Impact Group. 2018. Behind the meter PV solar funding guidebook. www.frontierimpact.com.au/external-resources

Fulcher, Jonathan and Elizabeth Harvey 2020. Legislation Update: New year, new review of the CATSI Act. Accessed May 12, 2020. www.hopgoodganim.com.au/page/knowledge-centre/legislation-update/new-year-new-review-of-the-catsi-act

Gammage, Bill. et al. 2021. *Country: future fire, future farming.* Port Melbourne, Victoria: Thames & Hudson.

Gibson, Graham, J.-K. 2008. "Diverse Economies: Performative practices for other worlds." *Progress in Human Geography* 32 (5): 613–632.

Gibson, Chanel Ann, Mehdi Aghaei Meybodi and Masud Behnia. 2014. "Investigation of a gas turbine CHP system under the carbon price in Australia considering natural gas and biogas fuels." *Applied Thermal Engineering* 68 (1–2): 26–35. https://doi.org/10.1016/j.applthermaleng.2014.04.002

Girardet, Herbert. 2008. *Cities, people, planet: Urban development and climate change.* 2nd ed. Chichester: John Wiley.

Global Power System Transformation Consortium. Accessed March 3, 2021. https://globalpst.org/

Goetzmann, William N. 2017. *Money changes everything: How finance made civilization possible.* New Jersey: Princeton University Press.

Goldthau, Andreas. 2014. "Rethinking the governance of energy infrastructure: Scale, decentralization and polycentrism." *Energy Research & Social Science* 2014 (1): 134–140. https://doi.org/10.1016/j.erss.2014.02.009

Granovskii, Mikhail, Ibrahim Dincer and Marc A. Rosen. 2006. "Economic and environmental comparison of conventional, hybrid, electric and hydrogen fuel cell vehicles." *Journal of Power Sources* 159 (2006): 1186–1193. https://doi.org/10.1016/j.jpowsour.2005.11.086

Grant, Eliza. et al. (2018) *The Handbook of Contemporary Indigenous Architecture.* 1st ed. 2018. [Online]. Singapore: Springer Singapore.

Guruswamy, L Lakshman. & Neville, E. 2016. *International energy and poverty: The emerging contours.* Abingdon, Oxon: Routledge.

Hallett, Steve and John Wright. 2011. *Life without oil: Why we must shift to a new energy future.* Amherst: Prometheus Books.

Harvey, Fiona. 2020. The Guardian, "China pledges to become carbon neutral before 2060." 23 September 2020. www.theguardian.com/environment/2020/sep/22/china-pledges-to-reach-carbon-neutrality-before-2060

Hawken, Paul. et al. (2010) *Natural capitalism: The next industrial revolution.* 10th anniversary edition. Abingdon, Oxon: Earthscan.

Helm, Dieter. 2015. *Natural capital valuing the planet.* New Haven: Yale University Press.

Hicks, Jarra. and Nicola Ison. 2012. 'What is community energy?' in *Home Energy Handbook*, edited by Allen Shepherd, Paul Allen and Peter Harper. Machynlleth: Centre for Alternative Technology.

Hicks, Jarra. and Nicola Ison. 2018. "An exploration of the boundaries of 'community' in community renewable energy projects: Navigating between motivations and context." *Energy Policy* 113: 523–534 https://doi.org/10.1016/j.enpol.2017.10.031

Hicks, Jarra. 2018. *Community power: Understanding the outcomes and impacts from community owned wind energy projects in small regional communities.* Sydney: University of New South Wales.

Hinchliffe, Sarah. 2012. "'Accounting' for intangibles: Taxation, intellectual property and the law … lost in translation?" *Intellectual Property Law Bulletin* 24 (4): 66–70.

Hollis, Leo. 2013. *Cities are good for you: The genius of the metropolis.* New York: Bloomsbury Press.

Homebiogas. n.d. "What is biogas? A beginner's guide." Accessed May 7, 2020. www.homebiogas.com/Blog/142/What_is_Biogas%7Cfq%7C_A_Beginners_Guide

Horackzec, Stan. 2021. "This two-way charger turns electric cars into a backup power source for your home." www.popsci.com/story/technology/dcbel-home-electric-car-charger/

Indigenous Business Association. 2018. Annual Report 2017–2018. www.iba.gov.au/wp-content/uploads/IBA-2017-18-AR-FINAL-WEB.pdf

Indigenous Business Australia. 2017. "Investor story: Manungurra Aboriginal Corporation." Youtube video, June 5, 2017. www.youtube.com/watch?v=HtSSixMQdVI&feature=share

Indigenous Business Australia. n.d. "Case study – Manungurra Aboriginal Corporation." Accessed May 24, 2019. www.iba.gov.au/investments/case-studies/case-study-manungurra-aboriginal-corporation/

Indigenous Business Australia. n.d. "Homepage." Accessed May 2020. www.iba.gov.au/

International Energy Agency. "Clean Coal Technologies" www.iea.org/reports/clean-coal-technologies

Janke, Terri. *True Tracks.* NewSouth Publisher, Sydney. 2021.

Johns, Gary Thomas. 2011. *Aboriginal self-determination: The Whiteman's dream.* Ballan: Connor Court Publishing.

Jock Collins and Branka Krivokapic-Skoko et al. 2015. "Community–owned Indigenous businesses and Indigenous co-operatives across remote, regional and urban Australia: The role and overall performance". Submission to the Senate Enquiry into Cooperatives, Mutual and member owned Firms.

Kanagawa, Makoto and Toshihiko Nakata. 2008. "Assessment of access to electricity and the socio-economic impacts in rural areas of developing countries." *Energy Policy* 36 (6): 2016–2029. https://doi.org/10.1016/j.enpol.2008.01.041

Keim, Stephen and Sean Reidy. 2015. Free, prior and informed consent: A just accommodation demands no less. www.academia.edu/19620322/Free_Prior_and_Informed_Consent_A_Just_Accommodation_Demands_No_Less

Keim, Stephen. 2018. "Book review: A World of Three Zeroes" at https://independent.academia.edu/StephenKeim

Keim, Stephen. 2018. "Book review: Janke, T., True Tracks" at https://independent.academia.edu/StephenKeim

Kemp, Adrian and Martin Chow. 2018. Open consumer energy data: Applying a Consumer Data Right to the energy sector. www.coagenergycouncil.gov.au/sites/prod.energycouncil/files/publications/documents/Consumer%20Energy%20Data%20final%20report.pdf

Kennedy, Jade et al. 2016. *A beginner's guide to incorporating Aboriginal perspectives into engineering curricula.* Australia: Engineering across cultures. http://indigenousengineering.org.au/wp-content/uploads/2016/07/eac-v4-small2.pdf pp. 17–28.

Kenny, Susan and Phil Connors. 2017. *Developing communities for the future*. 5th ed. South Melbourne: Cengage Learning Australia.

Keohane, Nathaniel O. and Sheila M. Olmstead. 2017. *Markets and the environment*. 2nd ed. Washington: Island Press.

King, Sarah, Naomi J. Boxall and Anand I. Bhatt. 2018. *Lithium battery recycling in Australia: Current status and opportunities for developing a new industry*. https://doi.org/10.25919/5b69ec381e06c

Klimenko, Svetlana and Eccles, Robert. 2019. "The Investor Revolution" Harvard Business Review, May-June Issue 2019.

KPMG. 2021. "The beginning of the ESG Regulatory Journey." Accessed July 30, 2021. https://home.kpmg/xx/en/home/insights/2020/05/beginning-of-esg-regulatory-journey.html

Lane, Taryn Adele. 2017. Investigating integrated systems of European zero net energy villages. http://c4ce.net.au/wp-content/uploads/2017/10/Churchill_Version-2_small.pdf

Langton, Marcia, Odette Mazel and Lisa Palmer. 2006. "The 'spirit' of the thing: The boundaries of Aboriginal economic relations at Australian common law." *The Australian Journal of Anthropology* 17 (3): 307–321. https://doi.org/10.1111/j.1835-9310.2006.tb00066.x

Laslett, Dean, Chris Creagh and Philip Jennings. 2014. "A method for generating synthetic hourly solar radiation data for any location in the south west of Western Australia, in a world wide web page." *Renewable Energy*. 68 (C): 87–102. https://doi.org/10.1016/j.renene.2014.01.015

Lazaroiu, George Cristian and Mariacristina Roscia. 2012. "Definition methodology for the smart cities model." *Energy* 47 (1): 326–332. https://doi.org/10.1016/j.energy.2012.09.028

Leon-Garcia Alberto, Radim Lenort, David Holman, David Stas, Veronika Krutilova, Pavel Wicher, Dagmar Cagáňová, Daniela Špirková, Julius Golej, and Kim Nguyen. 2016. *Smart City 360° First EAI International Summit, Smart City 360°, Bratislava, Slovakia and Toronto, Canada, October 13–26, 2015. Revised Selected Papers*. Cham: Springer International Publishing. https://doi.org/10.1007/978-3-319-33681-7

Lewis, Michael. 2018. *The fifth risk*. New York: W.W. Norton & Company.

Li, Yan, Eugenia Kalnay, Safa Motesharrei, Jorge Rivas, Fred Kucharski, Daniel Kirk-Davidoff, Eviatar Bach and Ning Zend. 2018. "Climate model shows large-scale wind and solar farms in the Sahara increase rain and vegetation." *Science* 361 (6406): 1019–1022. https://doi.org/10.1126/science.aar5629

Linklaters. 2019. Powering the future: Commercial opportunities and legal developments across the EV batteries lifecycle. www.allens.com.au/globalassets/pdfs/campaigns/linklaters-powering-the-future.pdf

Lou, Xian Fang, Jaya Nair and Goen Ho. 2013. Potential for energy generation from anaerobic digestion of food waste in Australia. *Waste Management & Research* 31 (3): 283–294. https://doi.org/10.1177/0734242X12474334

Lovelock, James. 1987. *Gaia: A new look at life on Earth*. Oxford: Oxford University Press.

Lovelock, James. 2006. *The revenge of Gaia*. London: Penguin.

Lukasiewicz, Anna, Stephen Dovers, Libby Robin, Jennifer McKay, Steven Schilizzi and Sonia Graham. 2017. *Natural resources and environmental justice: Australian perspectives*. Melbourne: CSIRO Publishing. https://doi.org/10.1071/9781486306381

Martin, Sarah. 2016. *Bush Heritage Australia: Restoring nature step by step*. Sydney: New South Books.

Matich, Blake. 2019. "Islands continue to turn to solar PV" *PV magazine*, November 19 2020. www.pv-magazine-australia.com/2019/11/19/islands-continue-to-turn-to-solar-pv/

Maisch, Marija. 2020. "AEMO: Lessons learned about VPP market participation" 28 March 2020. www.pv-magazine-australia.com/2020/03/28/aemo-lessons-learned-about-vpp-market-participation/

Mazengarb, Michael. 2021. "Australia's Martin Green awarded prestigious Japan Prize for work as 'Father of solar PV'" Renew Economy. Accessed March 4, 2021. https://reneweconomy.com.au/australias-martin-green-awarded-prestigious-japan-prize-for-work-as-father-of-solar-pv/

McDonald, Paula K. 2003. "Mapping patterns and perceptions of maternal labour force participation: Influences, trade-offs and policy implications." PhD thesis, Queensland University of Technology. https://eprints.qut.edu.au/15821/

McDonough, William and Michael Braungart. 2002. *Cradle to cradle: Remaking the way we make things*. New York: North Point Press.

McNeill, John Robert. 2000. *Something new under the sun: An environmental history of the twentieth century world*. New York: W.W. Norton.

Meinzen-Dick, Ruth. 2019. "Empowering Africa's women farmers" *The Optimist*, October 2 2019. www.gatesfoundation.org/TheOptimist/Articles/women-farmers-africa-gender-equality-agriculture-by-ruth-meinzen-dick

Mihyo, Paschal. and Mukuna, Truphena. (2015) *The gender-energy nexus in Eastern and Southern Africa*. Addis Ababa, Ethiopia: Organisation for Social Science Research in Eastern and Southern Africa OSSREA.

Munungurra Aboriginal Corporation. n.d. "Northern Territory development projects. Accessed April 29, 2020. https://manungurra.com.au/development/

Myers, Matthew, David Wolford, Danny Spina, Norman Prokop, Michael Krasowski, David Parker, Justin Cassidy, et al. 2016. "NASA Glenn Research Centre solar cell experiment onboard the International Space Station." *2016 IEEE 43rd Photovoltaic Specialists Conference (PVSC)*. Portland: IEEE. https://doi.org/10.1109/PVSC.2016.7750117

National Farmers' Federation. 2018. "NFF seeks action on high electricity prices." Accessed October 2018. www.nff.org.au/read/6169/nff-seeks-action-on-high-electricity.html

National Indigenous Australian Agency. n.d. "Grants and funding." Accessed May 2020. www.niaa.gov.au/indigenous-affairs/grants-and-funding

Nehme, Marina. 2014. "A comparison of the internal governance rules of Indigenous corporations: Before and after the introduction of the Corporations (Aboriginal and Torres Strait Islander) Act 2006." *Australian Journal of Corporate Law* 29 (1): 71–200. https://papers.ssrn.com/sol3/papers.cfm?abstract_id=3024549#

Nelson, Tim., Reid, Cameron. and Judith. McNeill. 2014. "Energy-only markets and renewable energy targets; complementary policy or policy collision?" AGL Applied Economic and Policy Research Working Paper No. 43.

Neto, Isabel and Anthony Maxwell. 2017. Austrade: Pacific Renewable Energy Project opportunities. www.pcreee.org/publication/austrade-pacific-renewable-energy-project-opportunities-webinar

Nettheim, Garth. 2000. "The re-recognition process for native title representative bodies: Pilbara aboriginal land council aboriginal corporation inc vs Minister for aboriginal and torres strait islander affairs." *Indigenous Law Bulletin* 5 (4): 77. www5.austlii.edu.au/au/journals/IndigLawB/2000/77.html

Nguyen, Kim, Jack J. Katzfey, John Riedl and Alberto Troccoli. 2017. "Potential impacts of solar arrays on regional climate and on array efficiency." *International Journal of Climatology* 37 (11): 4053–4064. https://doi.org/10.1002/joc.4995

Northern Territory Government. n.d. "Projects" Accessed April 4, 2019. https://bushtel.nt.gov.au/#!/projects

O'Shaughnessy, John. 2007. "Book reviews: Case studies and theory development in the social sciences: Alexander L. George and Andrew Bennett Cambridge, MA: MIT Press, 2005." *Journal of Macro Marketing* 27 (3): 320–323. https://doi.org/10.1177/0276146707305480

Ondraczek Janosch. 2014. "Are we there yet? Improving solar PV economics and power planning in developing countries: The case of Kenya", *Renewable and Sustainable Energy Reviews* 30: 604–615. https://doi.org/10.1016/j.rser.2013.10.010

Organisation for Economic Co-Operation and Development. 2010. Guidance on sustainability impact assessment. https://doi.org/10.1787/9789264086913-en

Parkinson, Giles. 2015. *Hazelwood owner promises no new coal fired power stations.* Renew Economy.

Parkinson, Giles. 2020. "Big new solar farm in NSW begins production, on schedule for a change" at https://reneweconomy.com.au/big-new-solar-farm-in-nsw-begins-production-on-schedule-for-a-change-55369/ accessed 10 June 2022.

Pascoe, Bruce. 2014. *Dark Emu: Black seeds agriculture or accident?* Sydney: Magabala Books.

Pedersen, Eja, Frits Van Den Berg, Roel Bakker and Jelte Bouma. 2009. "Response to noise from modern wind farms in The Netherlands." *Journal of the Acoustical Society of America* 126 (2): 634–643. https://doi.org/10.1121/1.3160293

Pernick, Ron and Clint Wilder. 2007. *The clean tech revolution: The next big growth and investment opportunity*. Pymble: HarperCollins.

Pernick, Ron and Clint Wilder. 2014. *Summary: The clean tech revolution: The next big growth and investment opportunity*. Cork: Primento Digital.

Pert, Petina et al 2013. *Indigenous Land Management in Australia: Extent, scope, diversity, barriers and success factors.* Cairns: CSIRO Ecosystem Sciences.

Pickin, Joe. and Jenny. Trinh. 2019. Australian Federal Government, Data on exports of Australian wastes 2018–2019, (version 2); 236.

Potter, Ben. 2021. "Green Bonds to hit $850b in 2021" Australian Financial Review. Accessed 18 March 2021. www.afr.com/companies/financial-services/green-bonds-to-hit-850b-in-2021-20210205-p56zzn

Powerlink. 2018. Transmission annual planning report 2018. www.transgrid.com.au/news-views/publications/Documents/Transmission%20Annual%20Planning%20Report%202018%20TransGrid.pdf

Preiss, Benjamin. 2019. "Testing begins for first offshore wind farm in Australia." *The Age*, November 10, 2019. www.theage.com.au/national/victoria/testing-begins-for-first-offshore-wind-farm-in-australia-20191110-p53970.html

Principles for Responsible Investment. "Delivering sustainable social infrastructure in Australia." Accessed July 30, 2021. www.unpri.org/infrastructure/delivering-sustainable-social-infrastructure-in-australia/7832.article

Putill, James. 2022. "EV chargers for V2G and V2H to arrive in Australia within weeks, after long delays." www.abc.net.au/news/science/2022-02-14/electric-vehicle-first-ev-chargers-v2g-v2h-to-arrive-australia/100811130

PWC, Strategy and Business News. "Investors and Asset Managers are Chasing a New Shade of Green," Accessed May 21, 2020. www.strategy-business.com/blog/Investors-and-asset-managers-are-chasing-a-new-shade-of-green?gko=2a745%C2%A0%C2%A0

Quality Circular Polymers. 2017. Suez and Lyondellbasell to boost production of high-quality recycled plastics in Europe. www.qcpolymers.com/wp-content/uploads/2019/04/20171127-Press-release-QCP.pdf

Queensland Government, Department of Environment and Science, Parks and Forests. "Fire Management". 2017–2022. Accessed February 26, 2021. https://parks.des.qld.gov.au/management/programs/fire-management

Queensland Government, Department of Natural Resources, Mines and Energy. 2020. "Solar for remote communities." Accessed August 10, 2020. www.dnrme.qld.gov.au/energy/initiatives/solar-remote-communities

Queensland Government, Department of Natural Resources and Mines. 2020. Leasing Aboriginal Deed of Grant in Trust land. www.dnrme.qld.gov.au/__data/assets/pdf_file/0016/107017/leasing-aboriginal-deed-grant-tust-land.pdf

Queensland Treasury. 2020. "Improving rehabilitation and financial assurance outcomes in the resources sector." Accessed February 3, 2020. www.treasury.qld.gov.au/programs-and-policies/improving-rehabilitation-financial-assurance-outcomes-resources-sector/

Quinn, Tom. 2015. "The Next Boom: A surprise new hope for Australia's economy." *Future Business Council.* www.fdocuments.net/document/the-next-boom-a-surprise-hope-for-australias-economy.html?page=4

RAG-Stiftung. 2018. Zeiten Wenden Annual Report 2018. www.rag-stiftung.de/fileadmin/user_upload/Publikationen/RAG-Stiftung_GB_2018_EN.pdf

Rahman, Aminur. 2001. *Women and microcredit in rural Bangladesh: An anthropological study of the rhetoric and realities of Grameen Bank lending.* Boulder: Westview Press. https://doi.org/10.4324/9780429503023

Rawls, John. 1971. *A theory of justice.* Cambridge: Belknap Press.

Recycling Magazine. "NGOs publish policy brief on recycled carbon fuels." Accessed June17, 2020. www.recycling-magazine.com/2020/06/17/ngos-publish-policy-brief-on-recycled-carbon-fuels/

Redback Technologies. 2020. "Responses to SA Government Regulatory Changes for Smarter Homes,"Version 1, 9th of September, 2020.

Repower Shoalhaven. n.d. "Repower One." Accessed April 8, 2020. www.repower.net.au/repower-one.html

Reserve Bank of Australia. "Renewable Energy Investment in Australia," Accessed March 19, 2020. www.rba.gov.au/publications/bulletin/2020/mar/renewable-energy-investment-in-australia.html

Reserve Bank of India. 2020. "RBI Bulletin" at www.rbi.org.in/scripts/BS_ViewBulletin.aspx?Id=19775

Roberts, Elisabeth, David Beel, Lorna Philip and Leanne Townsend. 2017. "Rural resilience in a digital society: Editorial." *Journal of Rural Studies* 54 (C): 355–359. https://doi.org/10.1016/j.jrurstud.2017.06.010

Robinson, Lesley. "'I think people are prepared to listen': calls for more cultural burning in Southern Queensland after 2019 fire season," 4 June 2020. Accessed February 26, 2021. ABC News www.abc.net.au/news/2020-06-04/more-cultural-burning-queensland-bushfire-victor-steffensen/12317196

Rodwin, Lloyd and Robert M. Hollister. 1984. *Cities of the mind: Images and themes of the city in the social sciences.* New York: Plenum Press.

Rose, Carol M. 2007. "The moral subject of property." *William and Mary Law Review* 48 (5): 1897–2926.

Russell, Graeme. 2021. The First Worimi Story. www.youtube.com/watch?v=ZI9MvU_GxdM

Rystad Energy. 2018. "Current pace of easy sales in Norway sets the stage for rapid future vehicle fleet electrification." Accessed April 2020. www.rystadenergy.com/newsevents/news/press-releases/current-pace-of-ev-sales-in-norway-sets-the-stage-for-rapid-future-vehicle-fleet-electrification/

SBS News. 2020. "Hopes microgrid funds will help Indigenous communities move to renewables." Accessed July 2020. www.sbs.com.au/news/hopes-microgrid-funds-will-help-indigenous-communities-move-to-renewables

Sekisui House. 2018. Sustainability report: Focused on creating shared value. www.sekisuihouse.com.au/getmedia/ac808991-3ad9-471c-84b9-50e54287a83f/sustainability-report-2018-a3.pdf.aspx

Sen, Amartya. 2001. *Development as freedom*. Oxford: Oxford University Press.

Shere, Lawrence, Siddharth Trivedi, Samuel Roberts, Adriano Sciacovelli and Yulong Ding. 2019. "Synthesis and characterization of thermochemical storage material combining porous zeolite and inorganic salts." *Heat Transfer Engineering* 40 (13–24): 1176–1181. https://doi.org/10.1080/01457632.2018.1457266

Singer, Joseph William. 2009. "Democratic estates: Property law in a free and democratic society." *Cornell Law Review* 94 (4): 1009–2062.

Solar Citizens. 2016. Homegrown power plan. www.getup.org.au/campaigns/renewable-energy/homegrown-power-plan/homegrown-power-plan

Solar Hybrids. n.d. "Australian Government: Outback Power Program." Accessed April 13, 2020. www.solarhybrids.com.au/case-study/australian-government-outback-power-program/

Sovacool, Benjamin K. 2014. "What are we doing here? Analyzing fifteen years of energy scholarship and proposing a social science research agenda." *Energy Research & Social Science* 1 (C): 1–29. https://doi.org/10.1016/j.erss.2014.02.003

Sovacool, Benjamin. K. 2012. "Design principles for renewable energy programs in developing countries." *Energy & Environmental Science* 5 (11): 9157–9162. https://doi.org/10.1039/c2ee22468b

Sovacool, Benjamin. 2012. "New partnerships and business models for facilitating energy access." *Energy Policy* 47: 48–55.

Sovacool, Benjamin. K., Roman V. Sidortsov and Benjamin R. Jones. 2014. *Energy security, equality and justice*. Abingdon: Routledge.

Spark Infrastructure. "Investor Centre," Accessed April 17, 2020. www.sparkinfrastructure.com/investor-centre/tax-information

Star of the South. n.d. "Star of the South Project: Australia's first proposed offshore wind farm." Accessed May 19, 2020. www.starofthesouth.com.au/

Sustainability Victoria. 2018. Australia's first lithium battery recycling plant opens. www.sustainability.vic.gov.au/About-us/Latest-news/2018/04/26/04/57/Australias-first-lithium-battery-recycling-plant-opens

Sveiby, Karl-Erik. 2009. "Aboriginal principles for sustainable development as told in traditional law stories." *Sustainable development (Bradford, West Yorkshire, England). [Online]* 17 (6): 341–356.

Swaback, Vernon D. 2007. *Creating value: Smart development and green design*. Washington: Urban Land Institute.

Talen, Emily. 2013. *Charter of the new urbanism: Congress for the new urbanism*. 2nd ed. New York: McGraw-Hill Education.

TasNetworks. 2017. TasNetworks Submission to Finkel Review into the Future Security of the National Electricity Market. www.environment.gov.au/submissions/nem-review/tasnetworks.pdf

Terrain Solar. Home page. Accessed June 10, 2022. www.terrainsolar.com/

The Australian Institute. 2020. Seminar 20 August 2020, Speaker Christina Bu, CEP of the Norwegian Electric Vehicle Association, and Gile Parkinson, editor of the Renew Economy.

The Queensland University of Technology. 2020. Campus Information Notice, (3 August 2020) "QUT PV solar seven days".

Toksoz, Mina. 2014. *The Economist guide to country risk*. London: Profile Books

Toohey, Paul. 2014. "Bess Nungarrayi Price has lost 10 siblings and talks about life in an Aboriginal town camp." *NTNews*, May 9, 2014. www.ntnews.com.au/news/national/bess-nungarrayi-price-has-lost-10-siblings-and-talks-about-life-in-an-aboriginal-town-camp/news-story/c796223c05aa582c15d1e974b096bd17

Townsend, Anthony. 2013. *Smart cities: Big data, civic hackers, and the quest for a new utopia.* New York: W.W. Norton & Company.

Tran, Tran and Claire Stacey. 2016. "Wearing two hats: The conflicting governance roles of Native Title corporations and community/shire councils in remote Aboriginal and Torres Strait Islander communities." *Land, Rights, Laws: Issues of Native Title* 6 (4): 1–20. https://aiatsis.gov.au/research/research-themes/native-title-and-traditional-ownership/publications/issues-paper

Transgrid. n.d. "EnergyConnect" Accessed March 12, 2020. www.transgrid.com.au/what-we-do/projects/current-projects/SANSWInterconnector

U.S. Department of Commerce, National Oceanic and Atmospheric Administration. n.d. "Climate." Accessed November 25, 2019. www.noaa.gov/climate

UK Committee on Climate Change. 2018. Hydrogen in a low-carbon economy. www.theccc.org.uk/publication/hydrogen-in-a-low-carbon-economy/

UK Office for National Statistics. 2019. "UK Natural capital accounts: 2019." www.ons.gov.uk/economy/environmentalaccounts/bulletins/uknaturalcapitalaccounts/2019

United Kingdom Office of the Information Commissioner. 2018. Investigation into the use of data analytics in political campaigns: Investigation update. https://ico.org.uk/media/action-weve-taken/2259371/investigation-into-data-analytics-for-political-purposes-update.pdf

United Nations Environment Program. 2008. "Renewable Energy, Generating power, jobs and development."

United Nations, Intergovernmental Panel on Climate Change. 2015. Fifth assessment report. www.ipcc.ch/assessment-report/ar5/

United Nations, Intergovernmental Panel on Climate Change. 2018. Fifth assessment report: Summary for policymakers. www.ipcc.ch/site/assets/uploads/2018/02/ipcc_wg3_ar5_summary-for-policymakers.pdf

United Nations, Intergovernmental Panel on Climate Change. 2018. Global warming of 1.5°C: An IPCC Special Report. www.ipcc.ch/site/assets/uploads/sites/2/2019/06/SR15_Full_Report_Low_Res.pdf https://www.ipcc.ch/report/ar5/wg3/

United Nations. 1987. Report of the World Commission on Environment and Development: Our common future. https://sustainabledevelopment.un.org/content/documents/5987our-common-future.pdf

United Nations. 1997. Kyoto Protocol to the United Nations Framework Convention on Climate Change. https://unfccc.int/documents/2409

United Nations. 2015. Paris Agreement: United Nations Convention on Climate Change. https://unfccc.int/process/conferences/pastconferences/paris-climate-change-conference-november-2015/paris-agreement

United Nations. "Sustainable Development Goals." July 30, 2021. www.undp.org/sustainable-development-goals

University of Queensland Sustainability Office. n.d. "Warwick Solar Farm" Accessed May 24, 2019. https://sustainability.uq.edu.au/projects/renewable-energy/warwick-solar-farm; see also https://stories.uq.edu.au/news/2020/warwick-solar-farm-powers-uq-100-cent-renewable/index.html

Uralla Case Study. Zeronet Energy Town at https://z-net.org.au/wp-content/uploads/2015/10/ZNET-CaseStudy_Uralla_v18.pdf

Van Eygen, Emile. et al. 2018. "Circular economy of plastic packaging: Current practice and perspectives in Austria." *Waste management (Elmsford). [Online]* 7255–7264.

Victoria Government, Department of Environment, Land, Water and Planning. 2017. "Certificate of electrical safety." Accessed September 9, 2019. www.energy.vic.gov.au/renewable-energy/victorian-feed-in-tariff/whats-involved-in-going-solar/paperwork-required-for-solar/certificate-of-electrical-safety

Victoria Government, Department of Jobs, Precincts and Regions. 2015, Guide to community-owned renewable energy for Victorians. www.energy.vic.gov.au/__data/assets/pdf_file/0030/57945/Community-Energy-Projects-Guidelines-Booket-A4_-WEB.pdf

Vidot, Anna. 2017. "Farmers call for comprehensive energy strategy as power prices increase more than 100 per cent." *ABC Rural News,* February 13, 2017. www.abc.net.au/news/rural/2017-02-13/farmers-push-to-rein-in-power-prices/8266022

Vollan, Bjorn and Elinor Ostrom. 2010. "Cooperation and the commons." *Science* 330 (6006): 923–924. https://doi.org/10.1126/science.1198349

Vorrath, Sophie. 2015. "Molten salt storage for rooftop solar? SA invention wins Eureka prize" https://reneweconomy.com.au/molten-salt-storage-for-rooftop-solar-sa-invention-wins-eureka-prize-99882/

Vorrath, Sophie. 2016. "Australian company buys 50% stake in "game-changing" graphene battery storage technology." https://reneweconomy.com.au/australian-company-buys-50-stake-in-game-changing-graphene-battery-storage-technology-83796/

Vorrath, Sophie. 2020. "Off-grid battery to shift NT Titjikala community to solar only, by day." https://onestepoffthegrid.com.au/off-grid-battery-to-shift-nt-titjikala-community-to-solar-only-by-day/

Walters, Adam. 2016. The hole truth: The mess coal companies plan to leave in NSW. http://downloads.erinsights.com/reports/the_hole_truth_LR.pdf

Watson, Richard, Marie-Claude Boudreau and Adela J. Chen. 2010. "Information systems and environmentally sustainable development: Energy informatics and new directions for the community." *MIS Quarterly* 34 (1): 23–38. https://doi.org/10.2307/20721413

Way, Julie. 2008. "Storing the Sun: Molten Salt Provides Highly Efficient Thermal Storage." www.renewableenergyworld.com/storage/storing-the-sun-molten-salt-provides-highly-efficient-thermal-storage-52873/

Weiss, Edith Brown. 1992. *Environmental change and international law: New challenges and dimensions.* Tokyo: United Nations University Press.

Weizmann Institute of Science. 1998 "Quantum theory demonstrated: Observation affects reality." *ScienceDaily.* www.sciencedaily.com/releases/1998/02/980227055013

Wynn, Gerard and Javier Julve. 2016. A foundation-based framework for phasing out German lignite in Lausitz. https://ieefa.org/wp-content/uploads/2016/09/A-Foundation-Based-Framework-for-Phasing-Out-German-Lignite-in-Lausitz_September2016.pdf

Xiaojing, Sun. and Wood Mackenzie, "Solar Technology Got Cheaper and Better in the 2010s. Now What?" 17 December 2019. Accessed 5 March 2021. www.greentechmedia.com/articles/read/solar-pv-has-become-cheaper-and-better-in-the-2010s-now-what

Yang, Hao, Santhakumar Kannappan, Amaresh S. Pandian, Jae-Hyung Jang, Yun Sung Lee, and Wu Lu. 2015. "Nanoporous graphene materials by low-temperature vacuum-assisted thermal process for electrochemical energy storage." *Journal of Power Sources* 284 (2015): 146–253. https://doi.org/10.1016/j.jpowsour.2015.03.015

Young, Robert. A. 2012. *Stewardship of the built environment: Sustainability, preservation, and reuse.* Washington: Island Press

Yunus, Muhammad. 2017. *A world of three zeroes: The new economics of zero poverty, zero unemployment, and zero carbon emissions.* Melbourne: Scribe Publications

Zurba, Melanie and Micaela Trimble. 2014. "Youth as the inheritors of collaboration: Crises and factors that influence participation of the next generation in natural resource management." *Environmental Science & Policy* 42: 78–87. https://doi.org/10.1016/j.envsci.2014.05.009

Cases

AGL Energy Ltd v Queensland Competition Authority & Anor (2009) QSC 90.
Anderson and Director-General, Department of Environment and Conservation (2006) 144 ZLGERA 43.
De Lacey & Anor v Kagara Pty Ltd (2009) QLC 77.
Donovan v Struber & Ors (2011) QLC 45 [13–26].
Elliot v Brisbane City Council (2002) QPELR 425.
Greenpeace Australia Ltd v Redbank Power Co Pty Ltd (1994) NSWLEC 178.
Hancock Coal Pty Ltd v Kelly & Ors (2013) QLC 9.
Hancock Coal Pty Ltd v Kelly & Ors and Department of Environment and Heritage Protection (No. 4) (2014) QLC 12.
Minors Oposa v Secretary of State of the Department of Environment and Natural Resources (1994) 33 ILM 173.
New Acland Coal Pty Ltd v Ashman & Ors and Chief Executive, Department of Environment and Heritage Protection (No. 4) (2017) QLC 24.
Origin Energy Electricity Ltd & Anor v Queensland Competition Authority & Anor (2012) QSC 414.
Origin Energy Retail Ltd v Queensland Competition Authority & Anor (2009) QSC 90.
Paltridge v District Council of Grant (2011) SAERDC 23.
St Helen's Area Landcare and Coastcare Group Inc v Break O'Day Council (2007) TASSC 15.
Telstra Corporation Limited v Pine Rivers Shire Council & Ors (2001) QPE 014.
Telstra Corporation Ltd v Hornsby Shire Council (2006) NSWLEC 133.
Wigness & Ors v Kingham, President of the Land Court of Qld & Ors (2018) QSC 20.
Yarmirr and Others v Northern Territory and Others (1998) 156 ALR 370.

Legislation

Aboriginal and Torres Strait Islander Act (Cth) 2015.
Clean Energy Legislation (Carbon Tax Repeal) Act (Cth) 2014.
Clean Energy Regulator Act (Cth) 2011.
Clean Energy Regulator Act (Cth) 2011.
Competition and Consumer Act (Cth) 2010.
*Corporations (Aboriginal and Torres Strait Islander) Amendment (Strengthening Governance and Transparency) Bill 2018 (*Bill*).*
*Corporations Act (*Cth) (2001).
Electricity Act (Qld) 1994.
Electricity Act (SA) 1996.
Electricity Act (WA) 1945.
Electricity and Other Legislation (Batteries and Premium Feed-in Tariff) Amendment Bill 2017, Explanatory Memorandum.
Electricity Industry Act (Vic) 2000.
Electricity Reform Act (NT).
Electricity Supply Act (NSW) 1995.
Environmental Protection Act (Qld) 1994.
Final Determination – Interim Reliability Instrument Guidelines Retailer Reliability Obligation July 2019; *National Electricity (South Australia) (Retailer Reliability Obligation) Amendment Bill 2018.*
National Electricity Law (SA).
National Electricity Rules (AEMC).

National Energy Retail Law Act (Qld) 2014.
National Gas (Capacity Trading and Auctions) Amendment Rule 2018; National Electricity Rule, Proposed Amendments 2018.
Privacy Act (Cth) 1988.
Privacy Amendment (Notifiable Data Breaches) Act (Cth) 2017.
Renewable Energy (Electricity) (Large-Scale Generation Shortfall Charge) Act 2000.
Renewable Energy (Electricity) (Small-Scale Technology Shortfall Charge) Act 2010.
Renewable Energy (Electricity) Act 2000.
Right to Information Act (Cth) 2009.
Sustainable Planning Act (Qld) 2009 (Repealed).
The Commonwealth Constitution of Australia.
The Commonwealth of Australia, House of Representatives, Treasury Laws Amendment (Consumer Data Right) Bill 2018 Explanatory Memorandum (Circulated by authority of the Hon Josh Frydenberg MP).
The Parliament of The Commonwealth Of Australia, House Of Representatives, Treasury Laws Amendment (Consumer Data Right) Bill 2018 Explanatory Memorandum (Circulated by authority of the Hon Josh Frydenberg MP).
The Planning Act (Qld) (2016).
Universal Declaration of Human Rights.

APPENDIX

TABLE A.1 Units of measurement

Unit	*Explanation*
W	A unit of measurement that calculates the rate of energy transfer
kW	1,000 W
kWh	A unit of energy equivalent to 1 kW of power transfer for 1 hour. This is the standard unit for pricing electricity
MW	1,000 kW
MWh	1,000 kWh
TW	1 trillion W
TWh	1 TWh = 3.6 PJ
MJ	1,000 J. 1 MJ = 0.278 kWh
GJ	1,000 MJ
PJ	1 million GJ

INDEX

Printed in the United States
by Baker & Taylor Publisher Services